Planetary Eating

Planetary Eating

The Hidden Links between Your Plate and
Our Cosmic Neighborhood

Gidon Eshel

THE MIT PRESS   CAMBRIDGE, MASSACHUSETTS LONDON, ENGLAND

The MIT Press
Massachusetts Institute of Technology
77 Massachusetts Avenue, Cambridge, MA 02139
mitpress.mit.edu

© 2025 Massachusetts Institute of Technology

All rights reserved. No part of this book may be used to train artificial intelligence systems or reproduced in any form by any electronic or mechanical means (including photocopying, recording, or information storage and retrieval) without permission in writing from the publisher.

The MIT Press would like to thank the anonymous peer reviewers who provided comments on drafts of this book. The generous work of academic experts is essential for establishing the authority and quality of our publications. We acknowledge with gratitude the contributions of these otherwise uncredited readers.

This book was set in Stone Serif and Avenir by Westchester Publishing Services. Printed and bound in the United States of America.

Library of Congress Cataloging-in-Publication Data

Names: Eshel, Gidon, 1958- author.
Title: Planetary eating : the hidden links between your plate and our cosmic neighborhood / Gidon Eshel.
Description: Cambridge, Massachusetts : The MIT Press, [2025] | Includes bibliographical references and index.
Identifiers: LCCN 2024056052 (print) | LCCN 2024056053 (ebook) | ISBN 9780262552141 (paperback) | ISBN 9780262382694 (pdf) | ISBN 9780262382700 (epub)
Subjects: LCSH: Food consumption–Moral and ethical aspects. | Agricultural ecology–Moral and ethical aspects. | Beef–Moral and ethical aspects.
Classification: LCC TX357. E83 2025 (print) | LCC TX357 (ebook) | DDC 630.2/77–dc23/eng/20250131
LC record available at https://lccn.loc.gov/2024056052
LC ebook record available at https://lccn.loc.gov/2024056053

10 9 8 7 6 5 4 3 2 1

EU product safety and compliance information contact is: mitp-eu-gpsr@mit.edu

To my family, in New York—Laura, Adam, Laila, and Luna—and Tel Aviv—Ima, Tami, Yossi, Yaeli, Shoham et al.—and in loving memory of Aba, Nimrod (who you will meet briefly shortly), and of Becky, my friend and almost tireless co-explorer of the Northeastern woods of fourteen years

Contents

List of Figures   ix

Acknowledgments   xi

Prologue   xv

1   Introduction   1

I   Preliminaries: Current Food Debates

2   Current Impacts of Diet and Agriculture on Health and the Environment   11

3   Of Beef and Public Confusion, Take I: Nutrition   23

4   Of Beef and Public Confusion, Take II: The Environment   41

5   An Alternative Look at the Public Confusion about the Environmental Impacts of Beef   73

6   Beef Production Using Presumably "Optimal" Grazing Practices   87

II   Fundamentals

7   A Brief Cosmic–Planetary Background   101

8   How Earth's Workings Shape Agriculture   119

9   From Sunshine to Food   151

10   Smooth Earth Operations: Reservoirs, Fluxes, Cycles   171

11   Bringing Planetary Eating Home: How to Shop, Cook, and Eat to Maximize Environmental and Nutritional Benefits   199

Bibliography   217

Index   267

List of Figures

7.1 A schematic representation of the dependence of nuclear binding energy per nucleon (proton or neutron in the nucleus) on nuclear mass. The maximum stability per nucleon region (masses 56–65, including nuclei of iron, nickel, copper) is highlighted by gray shading. The helium-4 nucleus is also shown. Its position, well above the general climb of binding energy with nuclear mass between hydrogen and the maximum stability region, indicates exceptional stability. The maximum stability region (gray shading) separates distinct energy extraction regimes. To the left of the shading (mass range 1–56), a great deal of energy liberated by fusing light nuclei into heavier ones. Conversely, to the right of this region (masses above about 60), smaller amounts of energy are liberated by splitting heavier nuclei into lighter ones.   107

8.1 Observed climatological (long-term mean over decades) air specific humidity between the surface (the plot's bottom) and 9 km above the surface, where atmospheric pressure is 300 milibars (mb), under a third of the surface pressure, where pressure attains its column maximum. Winter (December–February) and summer (June–August) are shown in solid and dash-dotted lines, respectively. In the plot, specific humidity data (in grams of water vapor per kg of air mixture) are averaged over 125°–70°W, 35°–50°N. The gray bands show height range, with the lowermost and uppermost bands spanning 500–1,000 and 8,500–9,000 m above sea level.   125

8.2 Panel a: Twelve day evolution of air "water vapor deficit" (see text for details) following the onset of 200 m per day subsidence that warms and dries the air by about 2°C and 0.6 g per kg per day. Time implicitly progresses from day 0, the lower left most square, to day 12, given by the upper rightmost square. Background gray shading helps visualize the passage of time, from the beginning in the lightest gray to the 12th day in the darkest gray. Integer multiples of the initial deficit on day 0 are indicated, culminated by almost 20-fold increases on day 12. Panel b: The evolution over the same 12 day period of an index of cattle thermal comfort, with time presented explicitly along the horizontal axis. 131

10.1 A schematic representation of a large scale land–ocean–atmosphere circulation pattern that arises, fundamentally, from the land–ocean thermal disparities and the obvious vastly higher availability of water to evaporate over the oceans than over most land masses. The text provides further details. 175

Acknowledgments

One of my favorite podcasts is the *New York Times* Book Review. Until recently, it was hosted by one of my literary heroes, the great, soft-spoken and gentle yet fiercely intellectual Pamela Paul, then the Review's editor. One week, Paul interviewed a woman who said something to the effect of "I just hope my reader will have half as much fun as I did writing the book."

Well, you are not going to get this sentiment from this author. Writing this book has taken a lot out of me, and anybody who was unfortunate enough to be around me when I wrote it. So my first acknowledgments go to my nuclear family, Laura, Adam, and Laila, and Luna the rescue dog, for their support, patience (well, mostly, but far more than is reasonable), and generosity. You guys are the bedrock of my existence, and I cannot even begin to coherently express just how much I admire and love you all.

With Laila (16), there is more. Not only is she a great walking partner—sometimes chatty and hilarious and at other time pensive and probing, often about the natural world—she has been my most trusted, tireless editor. While the "e" is lowercase—Laila is far too nice and kind to be a professional Editor—her input has been absolutely indispensable for this Middle Eastern immigrant, and without Laila, you wouldn't have to slog through any of this!

When, in the endless process of producing this book, I got the editorial nod to transition from the dreaded Word environment to the professional counterpart—LaTex—and faced a near catastrophe, our technically brilliant son Adam (23) wrote the neat, clean, and error-free python code that converted the Word references to BibTex, and saved the day. This was a huge relief for the author in me, and—far more importantly—a source of great pride for a technical Aba who recognizes with gratitude that his technical skills have now been decisively eclipsed by his young son's.

On many a writing day (and weeks and months), my only companion most of the time was Becky, our previous and now sadly no more ill-tempered old rescue dog. She and I walked together the forests near our rural home, saw hawks and bears and coyotes and bobcats and black vultures together, and had a great time. Back home, I would read her out loud some of the sentences I found particularly challenging, and she would repay me with endless demands that offered interminable opportunities for getting distracted.

Turning next to bipedals of the scientific persuasion, none has had a more positive substantive impact on this work, and indeed my scientific life in the last two decades, than my friend and colleague, the Great Ron Milo of the Weizmann Institute of Science in Israel. Ron is too smart, generous, nice, kind, and eager to help than anybody has the right to hope for, let alone expect. A one-of-a-kind scientist, the kind with a recurring timescale in the decades, and a lovely, lovely man, a good, devoted, loyal friend, a great hiking companion with an uncanny knack for raising the most piercing and ultimately revelatory observations of the tangible natural world in the most pedestrian of places. Ron's keen political senses have helped me navigate the many challenges of a life of publishing and facing reviewers, and has moderated my writing—which can sometime be a bit on the volcanic side of verbal expression—too many times to recount, always saving me from needless future complications and making the output better. Every scientist, no, make that every person needs a Ron, and I got one!

Several planets orbiting the bright star that Ron is—Alon Shepon, Avi Flamholz, Sam Lovat, and Elad Noor—have also been dear friends and helpful colleagues. They have helped me understand some biology, and they make science so fun, for which I am eternally grateful. Special thanks also go to another star in this constellation, CalTech's Rob Philips, who generously provided valuable feedback and encouragement when I desperately needed some.

My many conversations with Yohay Carmel of the Technion, Israel's premier technical school, are enthusiastically acknowledged. Yohay and I met in our distant early youth, introduced by our mothers, who were best friends in British-mandate Palestine, and remained good friends until Carmel's untimely passing. It has been a real joy to work and coauthor with such a deep ecological thinker and a good, kind friend as Yohay.

My old pals at Lamont and Chicago, David Archer, Ray Pierrehumbert, Shanan Peters, Munir Humayun, Mike Steckler, Richard Seager, Larry Grossman, Jerry Mcmanus, Maureen Raymo, and Andy Davis, as well as Yair Rosenthal at Rutgers, kindly and generously read parts of the manuscript, and helpfully opined. I also thank Itzik Mizrahi and Shaul Pollak for some very helpful microbiological adult supervision. While the remaining errors are surely mine, I greatly appreciate their selfless help.

My more recent colleague and friend, David Katz (now of DietIQ and formerly of the Yale School of Medicine) generously and carefully read the more nutritional science elements of this book, and weighed in invaluably, for which I am deeply thankful. Less specific, yet just as appreciated, is Michael Pollan's ongoing support, generosity, and class. While I am decidedly no Michael Pollan, and never will be, he made aspiring fun, and emotionally tractable.

I am indebted to my friends/colleagues Eli Tziperman, Steve Wofsy, Peter Huybers, Robin Wordsworth, and Dan Schrag at the Harvard Department of Earth and Planetary Sciences for their useful, pertinent comments, and for their continued, ongoing generosity and hospitality, which contributed importantly to this book. Early parts of this book were likewise written during my nice 2016/2017 Radcliffe Fellowship, which some unknown combination of the above plus Michael Pollan cooked up for me. I thank Radcliffe, and the above people, deeply for that illustrious opportunity, an experience that offered just the type of pampered, coddled existence I never had before or since. I thank Harvard deeply for all of that.

I am humbled and touched by the kind, generous support of all the scientists mentioned above, from systems biology, biophysics, epidemiology, and geophysics. It has been particularly gratifying to reconnect with the latter community, in which I grew up, scientifically speaking, but from which I had foolishly largely diverged over the years during which I developed the geophysics of food work. The eagerness to help of all of these very busy global leaders, and their wise input during the course of this project have been truly heartwarming to me.

My final thanks go to my editor at MIT, Beth Clevenger, who diligently midwifed this complex and often choppy birth. Thank you so much, Beth!

Prologue

On an unseasonably balmy November 2006 evening, my friend Pam Martin and I arrived at the Princeton campus. Earlier that year, we had published *Diet, Energy and Global Warming*,[178] which got us an invitation to speak at a Princeton event titled Food, Ethics and the Environment, convened by the philosopher Peter Singer.[551] I felt excited for my first entry into "The Big Time," but on edge; I am more at home among deckhands or farmers than among formal, tweed clad Princeton professors. It was also nerve racking to share the podium with such luminaries as Eric Schlosser (of *Fast Food Nation* fame), Michael Pollan, or Marion Nestle, incredibly tough acts to follow.

But Michael, Marion, and the Princeton Environmental Institute staff turned out to be warm and friendly, helping me gradually overcome my unease and enjoy giving my talk and listening to some unforgettable ones by others on the formidable speaker roster.

After the meeting, I returned, Cinderella-like, to my drab Chicago existence, and life appeared to gently regress to the familiar routine of teaching, research, and above all parenting our five-year-old Adam. But in retrospect, that meeting proved a real fork in the road, after which my previous and current paths rapidly diverged. Perhaps emboldened by the recognition our work received there, or maybe simply guided by an internal sense of the optimal confluence of what I could do best and what's most important, from 2006 on, I have worked primarily on how geophysics (the totality of Earth's interconnected physical, biological, and chemical processes) relate to agriculture.

This new path has differed from the typical academic career in many ways, good and bad. Dominating the negative is that I never enjoyed the resources and nurturing protection some scientists enjoy from their home institutions. If you read (and I *warmly* recommend you do) such classics as *The Story of Earth*[253] or *The Second Kind of Impossible*,[572] the authors' emotional affiliation with their home institutions, and the gratitude they feel toward the research and work support they have enjoyed there are palpable. Their experience couldn't have been more distinct from my Chicago experience. Instead, the new career path has thrust me into an intense multidimensional public debate. One aspect involves the tension between an idealistic, Luddite-like worldview favoring small-scale, "local"[120,633] agriculture that is presumed environmentally virtuous on account of low "food miles" (how far and by what means the food has traveled)[374] on the one hand, and the currently dominant alternative of massively intensive corporate mega-farms on the other. Another very dominant debate centers on beef eating as either the source of or the cure to all agricultural ills. In keeping with this polarity, my popular writing and posted interviews typically receive online comments from individuals that are roughly evenly split between the blindingly enlightened, zero doubt vegan activists, and angry self-appointed beef and big ag defenders. And when I publish papers that suggest that *in some circumstances* beef *may* have some productive roles to play (it does[175,182]), the tenor of the comments remains unchanged, but the camps neatly reverse, like Prussian troops in formation.

Such incuriosity rarely begets serious fact- and logic-based inquiry, yet it is sadly common in the public discourse on food, agriculture, and diet. In this book, the only standard I have used is a strict adherence to what the guiding science—sometimes foundational, sometimes applied—reveals, and a deference to veritable, robust observations. I am guided by the (possibly naive) notion that if you just highlight sufficiently clearly, in widely understood language, the foundational, logical or factual holes in some of the commonly held views, people will listen. In other words, that polarization is curable, and that I can help cure it. Immodest though these thoughts are, I can't shake them off. This book is my effort to reassert the supremacy of simple logic and basic science in the discussion on the simultaneous environmental and nutritional betterment of agriculture.

It is tempting to begin where the story of the encroachment of our appetites into our landscapes really begins, variably pegged to the green revolution, the industrial revolution, colonialism, the European pillaging of the Americas, or the dawn of agriculture in the Fertile Crescent. While clearly important, these are merely recent chapters in a story whose prologue unfolded over 4.5 billion years ago, as our solar system coalesced from clouds of interstellar dust, or earlier still, when that dust acquired its material attributes from earlier generations of long-gone stars.[85,332] Fundamentally, agriculture builds directly on these processes, which—through nuclear interactions that involve protons and neutrons in atomic nuclei—create the tangible materials around us, and determine their chemical properties, as discussed briefly in chapter 7. In turn, these shape agriculture, which is nothing if not controlled fluxes of energy and mass between various interconnected components of the natural world.

If this seems odd, think about work-performing muscles; are they not fueled by the calories in our food, which are stored solar energy, liberated by nuclear fusion in our little star, the sun? Or think of the most common agricultural action, irrigation, the redistribution of water whose mass is about 90% oxygen. Is the oxygen nucleus not four helium nuclei, fused together in stellar interiors? Or fertilization, distributing biologically available nutrients on fields for crops to use as needed. Where did the elements—nitrogen, phosphorous, among others—these nutrients comprise come from? Like oxygen, the daughter products of stellar nuclear fusion (more on that in chapter 7). So the presence of these substances near Earth's crust, where agriculture takes place, is an inherently planetary sequel to an inherently cosmochemical–astrophysical earlier story.

Every twist in these plots could have gone very differently (as described in chapter 7). For example, if the cosmic neighborhood now occupied by our solar system had been characterized by much lower dust concentrations, possibly for want of nearby exploded earlier stars to seed that neighborhood with celestial building blocks, the solar system would have had much lower abundances of life relevant elements, and possibly even too low a total mass to ignite a stellar furnace.

Or, if Earth had formed much closer to the sun, the high temperatures this proximity entails would have likely doomed Earth to hold much less water and nitrogen.[451,474,653] Or, if Earth were much smaller, its feeble gravitational field would have likely allowed much hydrogen to escape to space, again leaking water to space. Clearly, Earth, and agriculture on its surface, bear the unmistakable signatures of astrophysical–cosmochemical processes.[19] Because of the above and many other deep roots that connect agriculture to fundamental physics, chemistry, and allied basic sciences, to me the story of agriculture is not about Luddite longings to a bygone era of local, simple agriculture, or improbable promises for magically effortless environmental betterment by any one of the pronouncements routinely made in this space. Rather, the grammar and vocabulary of this story are basic scientific principles applicable to human–Earth interactions.

Consequently, this book synthesizes many observations of the natural world, most widely known for a long time. The novelty therefore stems from combining information gleaned from the agricultural sciences, farming practices, nutritional science, and the suite of natural sciences that collectively govern most of our experiencing of our environment: fluid dynamics on a rotating sphere, meteorology, oceanography, biogeochemistry, hydrology, among others, and—most importantly—the interactions of all of those with each other and with ecology. Such syntheses typically do not spontaneously occur. Instead, they often mirror—and are made possible by—the personal paths of the synthesizers. Let me therefore share with you a few highlights of my path.

I grew up on the ocean, the son of a ship captain. My parents, sister, and I shared a diminutive cabin on board an old, small ship. Because we mostly traveled northern Europe routes—traversing regularly the Bay of Biscay, English Channel, the North and Baltic seas, waterways that are second only to the Southern Ocean in fierceness[145]—our main deck, and at times even the bridge, were dwarfed by the seas around us.

For years my older sister Tami was the only fellow child I knew. When the weather cooperated (it rarely did), I worked with the deck crew. When the weather raged, and decks became off-limits, I had a perch all my own just behind the bridge windows on the ship's starboard (right) side. To my left was the indomitable captain Eshel, Aba to me, pacing back and forth along the narrow gap between the instrument panel behind and those windows ahead. Around Beaufort force 9 (this sea state scale was one of my closest childhood companions, and in force 9 even seasoned seamen are somewhat seasick), our existence became tenuous. I would watch our bow descending into the incoming trough with a grim splash, followed by a wall of green water and whitish-gray foam completely inundating the windows before me. Then, quiet. The ship, stern raised high and bow dipped low, was already pointing toward the abyss, as though the deed is done, the doers undone. The tense silence on the bridge yields the full audible landscape, magnifying manyfold the shriek of the wind against taut wires and the groans of deforming steel plates. A full eternity is condensed into several seconds (in inland waters) or much longer (in the open ocean) during which all eyes are firmly trained on the patch of ocean that now covers our bow and fo'c'sle deck. And then, tentatively, hesitantly, buoyancy prevails. The bow begins rising, and familiar objects gradually resurface. First the small foremast just behind the anchor winch. Then the bulwarks hugging the bow deck, decorated forlornly by dripping flotsam. And finally, the fo'c'sle watertight steel doors become visible once more. Exhale. Repeat. Sometimes for hours, at other times for days or more.

For Tami and me, schooling, such as it was, constituted an unusually lax one-woman Department of Education our mother ran on board, sea state permitting. With months at sea at a stretch, and neither a formal education nor any structured commitments to avert suffocating boredom, we became consummate curious self-educators.

From age five or so, the natural world I witnessed from the ship was what truly captivated me. Nature grabbed hold of me, and never let go. What turned the lazy green-brown waters of the sheltered Elbe, I wondered, into a gray tempest upon departing Antwerp into the North Sea? Why did the still Malmo fog give way to glassy ice as we crossed the Baltic sea southward, toward the Kiel canal and ultimately Hamburg? Why are the eastern Adriatic waters near Koper gin-clear, while approaching Venice, just 100 nautical miles (or 130 miles) away, is like floating on bad coffee? My initial processing of these observations was wholly thanks to my father. With zero formal schooling, he rose up the hosepipe, as such paths are described on ships, from galley boy to captain, in the rigorous, unforgiving traditional British maritime system, whose old-timers still mourned the loss of the lash as an educational tool. Under his watchful eye, and thanks to his innate pedagogical gifts, I began to cogently translate eyewitness observations into coherent mechanistic ideas. Decades later, working with two great mentors, Mark A. Cane at Columbia and Brian F. Farrell at Harvard, all these years of maritime observations gradually cohered into a unified story. Generously and rigorously nurtured by these two titans of modern fluid physics, I was finally able to turn basic observations of the natural world into meticulous, crisp scientific constructs that, given the breadth of my inquiry, required insights from numerous scientific disciplines.

After completing my PhD and moving on to research or faculty positions at Harvard, the Woods Hole Oceanographic Institute, the University of Chicago, and eventually Bard College, I taught myself ever more disciplines, and assimilated each into an ever-evolving, richer, and more nuanced and integrative view of Earth workings. This synthesis and the physics-based worldview it yielded remain my primary tools in research, and the interpretive lens used in this book.

But this book is not about the physics of hurricanes or cyclones; it's not even about aquaculture. Where does agriculture come in?

At age thirteen, I decided that I had taunted Poseidon enough. I went landlocked, and stumbled onto what would become a second formative experience. I bid my parents adieu, and moved to Kibbutz Mizra, in northern Israel, where I first saw a school from the inside. To me, kibbutz high school life, as a member of a semi-independent children principality, was as close to paradise as can be. You live with your classmates, two to four in a room, in a small house where grownups play minor and transient roles and can at most times be safely ignored. In this autonomous environment I enjoyed my first up-close and personal encounter with cattle, from which many cattle-focused years naturally flowed. First was the kibbutz dairy farm. (In case you're unfamiliar with the standard high school life of kibbutz kids, a day comprises school before lunchtime, followed by working in the part of the farm of one's choosing until dinner.) I had great mentors—generous, knowledgeable, tight-lipped men in their forties and fifties who lived and breathed dairy farming—and I was instantly hooked. The dairy herd became the centerpiece of my young self, all the more so during and following the 1973 Yom Kippur War (when all men were on reserve duty, and we kids ran the entire farming show ourselves for several months). Then at twenty-two, after mandatory military service, I returned to cattle, this time focused on beef rather than dairy, in Kfar Yehoshua and on the western slopes of the Golan Heights. Collectively, I've enjoyed fifteen years of hands-on cattle farming experience. There is no aspect of cattle farming that I haven't personally handled countless times.

During those years, I also completed a formal Cattle Herd Management course run by the Israeli Ministry of Agriculture. Designed expressly for practicing herd managers, this very intensive, rigorously scientific two-year residential course was taught by scientists from the Hebrew University and from governmental agricultural research institutes. It covered cattle physiology, veterinary medicine, animal and human nutrition, biochemistry, agricultural economics, and agrobusiness optimization. All of these farm-focused experiences deeply impacted and informed my worldview, and gave me intimate familiarity with the practical farming details that are so central in later chapters.

When I eventually became a student of Earth and its workings, I thought I'd left farming behind for good, just like I left the ocean behind me earlier. But these clean breaks were never to be.

The merger of seafaring, farming, and Earth's workings into the unified geophysics of food worldview that *Planetary Eating* summarizes took a while, as it became progressively clearer to me that while geophysics and farming appear incongruent, the purview of traditionally nonoverlapping scientific disciplines, this separation serves no party well, because farming, which strongly impacts Earth, is applied geophysics in disguise. While this realization was gradual, one memorable event spearheaded this unexpected merger.

In the summer of 1996, shortly after marrying Laura and moving to Cambridge for a postdoc position, we visited Israel. I was eager to show Laura some particularly meaningful places to me, to give her a context for my origins, my family, myself. (All perplexed her, for which you could hardly blame the woman ...)

After a week of family and maritime-themed spots, we turned our attention to the other key pillar of my existence, farming. We visited Mizra, and then friends in two kibbutzim in the Yarden basin, south of Lake Kineret. That trip proved revelatory—to me, paradoxically, despite the focus on Laura's interests. Now to understand what follows, you have to know this: Laura grew up in New York; farming was no more familiar to her than space exploration. But she gets people, and could tell that to understand these old friends of mine, their unfamiliar environment, life in an old Israeli kibbutz, and even her peculiar new husband, she needed to better understand farming. Farming, in those final days of the kibbutz movement, was to that community what Grand Central is to New York commuters: the origin of everything, to which everything eventually converges. So she asked. And asked, and asked some more.

As these conversations branched and deepened, earth sciences became ever more present in my answers, whether related to meteorology and climate, geophysical fluid dynamics, atmospheric thermodynamics, or surface hydrology. Eventually, as our odyssey unfolded, it became hard to miss that, when Laura posed farming questions, I answered in the language of geophysics. By the time we drove back to Tel Aviv and then flew back to Boston, I could miss it no more: my new world—the equations that govern Earth's fluid envelope—and the burning passion of my younger self—farming—were forming a novel union in my head. It also became clear why that was happening; just beneath the surface, farming and geophysics are mechanistically, physically inseparable. It is often said that you cannot meaningfully understand food without knowing how, when, where, by whom, and by what means it is produced. I find parts of this somewhat persuasive, while rejecting the notion that it's about getting to know your farmer's smile or her folksy persona. Rather, it is about the impossibility of meaningfully deciphering these whens, wheres, and hows of food production without invoking basic physics, thermodynamics, biology, and other pertinent sciences. Basic knowledge from each governs our interactions with Earth's moving parts—the land, atmosphere, or vegetation—and determines our survival and success.

Some food system analysts invoke economics to explain agriculture and our food system, with some modest success. Yet, at least to me, the explanations basic science offers are far much more powerful, naturally generalizable, and steeped with so much more elegance and aplomb. This is why modern food discourse stands to gain so much from a clear, thorough exploration of the geophysics of food, the scientifically inquisitive alternative to our current, often muddled food conversations. By reuniting agriculture with its scientific roots, from which it was severed by academic idiosyncrasies rather than logic, geophysics of food offers a currently unavailable level of deep, mechanistic understanding of food and its production, and tells this story with the unsurpassed rigor and intellectual elegance that science affords.

Introducing geophysics of food to sophisticated and curious yet scientifically untrained readers interested in food is what compelled me to write this book. In the following chapters, we will gradually cover what geophysics of food actually does, and how it looks and feels doing so. I will share key elements of this story and the myriad scientific constraints that help navigate the otherwise bewildering, hopelessly open-ended discussion about what matters for considering environmental impacts of our food choices and agricultural practices. If you give me your attention—and I grant you this book is probably not everybody's idea of a light read—it will become a lot easier to differentiate the serious elements of our food discourse from those that are best ignored.

Far from merely an intellectual exercise, the ability to successfully distinguish idle gratuitous assertions from sound facts or logical conjectures can prevent making nutritionally and environmentally damaging personal choices, and prevent supporting bad agricultural policies that undermine public health while promoting national scale waste of resources, effort, and goodwill. This book is premised on my belief that the current state of confusing affairs is neither necessary nor inevitable. Clarity is possible, *provided* one takes the trouble to gain a basic mechanistic appreciation of the relevant processes. In this book I therefore set out to reduce public confusion about food by distilling the scientific essence of agricultural human–Earth interactions into as condensed a synopsis as I could. Because the scope of this task is impossibly large, I had to curate example scenarios down to a few key ones that are of considerable societal import and that exemplify interesting, generally applicable scientific ideas that are still technically relatively accessible. My choices also reflect my taste, involving idiosyncratically selected vignettes with which I hope to illuminate much broader, more general truths. I ignore many of the "food fights" that have sadly come to almost define this science, but address a few that really matter, in which assertions that run particularly flamboyantly afoul of the governing science abound.

Throughout this work, my tools of choice are pertinent physics and related sciences that reveal the fundamental Earth processes agriculture perturbs and sometimes commandeers. If this scientific aspect gives you pause, fear not. My intended audience comprises interested, curious readers with no technical background, and I assume no earth science or physics background beyond typical high school levels. To be sure, there are challenging parts here. For example, in later chapters we address precipitation-generating processes (the processes that yield drizzle, downpour, or snow), or their stifling by subsidence (downward wind) whose impact on agriculture is unmistakably dominant. Understanding precipitation-generating processes requires exploring phase transitions—mostly evaporation and condensation—in the atmosphere and over land surfaces, which in turn, requires excursions into thermodynamics and stability. So there are Russian doll–like nested intellectual jaunts, each necessitated by a requisite concept or body of knowledge I just cannot reasonably assume all readers possess, but I summarize each at the end, allowing the perplexed to remount the moving train.

These are my modest tasks, and I am eager to embark on this odyssey with you, starting now.

Gidon Eshel
Rhinebeck, NY spring–summer 2024

1

Introduction

For its most recent half a million years, Earth has been quasi-regularly varying between glacial and interglacial conditions every roughly 100,000 years.[575] Following the so-called Last Glacial Maximum, some twenty thousand years ago, until about eleven thousand years ago, when the current geological epoch, the Holocene interglacial, took a firm and for now final hold, the most recent glacial-to-interglacial transition unfolded. Between the Last Glacial Maximum nadir and about nine thousand years ago, global mean annual temperatures arguably climbed by about 8°C (roughly 14°F).[637] The warming then continued much more slowly, adding no more than 0.5°C (1°F) by the mid Holocene (about six thousand years ago), before switching to our current rapid anthropogenic global warming.[202] (Paleo-thermometry—indirectly inferring past temperatures—is based on various proxies, notably one based on the ratio of stable oxygen isotopes in such geological samples as deep sea sediments or cored ancient ice. Why this works, or its governing physics are not essential here.) Most of our cultural and technological development unfolded in the relative warmth of this timespan.

The currently prevailing explanation of these glaciation events— widely known as ice ages—is that they track subtle variations in Earth's orbit around the sun and in the tilt and gyration of Earth's own axis of rotation. Because these orbital characteristics vary semiperiodically, glacial cycles are expected to be similarly semiperiodic, and successive events should closely track one another. That appeared true for the 3–4 millennia following the Last Glacial Maximum, with gradual warming more or less tracking both orbital changes and the warming that characterized corresponding phases of earlier glacial terminations and transitions into their respective subsequent interglacials.[504-506] But around seven thousand or so years ago, the current interglacial began misbehaving.[69] First, global mean surface temperature maintained unusually high interglacial warmth throughout the Holocene (the most recent interglacial), in marked deviation from earlier interglacials, in corresponding points of which warming slowed, stabilized briefly, and then reversed, with global temperatures accelerating back toward glacial chill. Atmospheric concentrations of two key greenhouse gases, carbon dioxide and methane (CO_2 and CH_4), also diverged.[78] (We infer past atmospheric composition from air bubbles trapped in cored polar ice sheets, crushing small individual ice samples, liberating the gas bubbles trapped in each, and measuring their composition, which is that of the atmosphere when the bubbles closed off and lost contact with the atmosphere. Apart from some technical subtleties, the age of the ice varies from approximately "now" at the surface to ever earlier down the core, with an aging rate determined by various age models whose mechanistic foundations do not matter here.) In earlier interglacials, atmospheric concentrations of CO_2 and CH_4 rose minimally right after the glacial termination but gradually sank back to much lower glacial levels throughout almost the entire interglacial. Holocene levels briefly followed suit in the early postglacial, steadily declining over 7–4 thousand years ago, as expected based on Earth's orbit variations. Then, however, this decline reversed,[266] with carbon dioxide and methane rising again, eventually switching to today's exponential rise.

In other words, as Earth has been recovering from its most recent glaciation, but *well* before the onset of widespread intensive agriculture or industrialization, it has been charting an unusual climatic path markedly distinct from its behavior during earlier glacial cycles. Instead of the canonical brief flickering of a post glacial termination warmth followed by a rapid descent right back to essentially glacial conditions, geologically recent climate has been stuck in interglacial conditions. To be clear, the exponential rise in atmospheric concentrations of greenhouse gases in the last two centuries or so mentioned earlier is unsurprising, being just what's expected given our emissions of these gases into the atmosphere. What *is* surprising is the rising[574] during several millennia *well before* the Industrial Revolution and the widespread proliferation of fossil fuel combustion that fueled it.

So, what is so special about our current interglacial? Why did Earth behave unexpectedly in the 6–7 millennia *prior to* the Industrial Revolution (before, say, 1750)?

Several mechanisms have been invoked to explain the processes responsible for the marked departure of the path the earth system has been following in the Holocene from that of earlier interglacials. Some, including quite persuasive ones,[556,603] focus mostly or solely on natural processes, suggesting little human impact. Another—deforestation by intentional fires[300]—*is* related to food and agriculture, but partially and indirectly (because some deforestation is for construction, tool-making, warmth, and other uses not directly related to food). But others train the spotlight firmly on us. Of those, my subjective favorite is the Ruddiman Hypothesis,[504] the idea that our dietary choices started impacting Earth's climate as early as 7,000 years ago.

The idea is Bill Ruddiman's and is well worth exploring. It sheds important, revealing light on our story, scientifically as well as on an individual level. As I write this, Bill is a very active emeritus professor at the University of Virginia, a recipient of the Lyell Medal (the flagship honor of the Geological Society of London), and a Fellow of all the Societies a geologist wants to be a fellow of. While I doubt Bill remembers me, he was my Climate Change professor in my trying first months as a graduate student at Columbia. A kibbutz high school dropout, my general education was at best sketchy. I always read a lot, which helped *some*, but my English was rudimentary; I understood maybe one in six words on a good day, with the experience reminiscent of my most consuming terror, drowning. Yet here I was, a beginning PhD student in one of the 2–3 best geophysics graduate programs in the world.

The technical courses, the bulk of my course load, were no problem; Lagrange multipliers or eigenvectors are the same at the Technion and Columbia. But grasping climate change—mostly words, most with many, twisting syllables—was quite a different matter. The course was taught by the formidable Wally Broecker—a student of forces of nature, who is himself one, and perhaps the greatest interpreter of climate change geochemistry ever—and the soft-spoken Bill Ruddiman. While Wally had little use for questions, and his skin was always straining to contain his singular energy, Bill was self-deprecating and approachable. And he was patient with me, and much more friendly and caring than I deserved. He took time after each class to answer my many questions, which were delivered choppily with a thick Hebrew accent, and he artfully translated things to me by dialing down his language eight notches. And when I (rarely) got it, he rewarded me with an asymmetrical, wry, perhaps even shy smile.

So when, a decade later, I began studying the planetary impacts of agriculture, I quickly became curious about the earliest global imprint of these impacts—the 1980s? the 1950s, maybe?

Being from the Middle East, I was well aware of early technologies with which humans impacted their environment. I knew well and admired greatly the terrace-based Palestinian agriculture, at the time essentially a replica of biblical agriculture. Roughly waist to chest high, terraces slowed soil robbing runoff, allowing precious red Terra Rosa soil to collect in their upslope pockets, producing locally flat soil surfaces on which agriculture became just barely possible. While terracing enabled rather meager cropping, and required unending back breaking maintenance, it kept the steep, erosion prone western slopes of the north–south Israeli/Palestinian mountainous backbone from becoming a sterile wasteland despite ample precipitation. Biblical archaeologists believe that Israelites first developed this terraced landscape in the mountains of today's Palestine and Israel some 3,300 years ago because the fierce Philistines, backed by unmatched iron-based warfare know-how, already reigned over the easier to cultivate coastal plain. Because Middle Eastern precipitation is modest, and capricious, these terraced landscapes relied heavily on small local water systems. Called *ein* in both Arabic and Hebrew, these are tunnels, several meters long and roughly one square meter in cross section, painstakingly dug into fractured hard dolomite or limestone layers overlaying thin blankets of soft, essentially impervious clay that kept water in the tortuous crevices of the hard rock above. Those carved tunnels enhanced seepage from water bearing rock layers above and directed the harvested water

INTRODUCTION   5

into small collection pools. When you read "A garden inclosed is ... my spouse; a spring shut up, a fountain sealed" in the *Song of Songs* (or *Song of Solomon* in the butchered English translation), you are reading a racy metaphor based on these life sustaining water systems. (Parts of the Old Testament are an agricultural almanac disguised as Hebrew poetry.)

I also knew of the qanat (see chapter 21 of reference 230), another ingeniously designed system of subterranean tunnels that harvest water seepage by following local water tables. I first encountered qanats—or fugaras, as they are known in Palestine and Israel—in the process of making sense of puzzling straight line segments tens to hundreds of meters long that were glaringly visible in aerial photographs of the arid flats north of Jericho. Rather than hard rock, like the ein systems, fugaras are dug into the loose, poorly consolidated soils of the Jordan River's flood plains. On closer examination of higher resolution photos, it became clear that these straight lines comprise anywhere from a handful to tens or more evenly spaced circular surface mounds. They were clearly made by humans, yet no walls, roads, or any other human made object one regularly sees in aerial photography were apparent. What were these mounds? They remained a mystery until Israeli archaeologist Shimon Dar taught me years later that they are the aboveground signature of the fugara system; the mounds served as a repository for excess soil and as breathing holes for the small children who maintained the tunnels in the face of ceaseless crumbling and clogging. Apart from the oppressive child labor aspect, qanats worked so well that their use propagated from their Persian birthplace throughout the ancient Near East.

But terraces, qanats, and such are landscape specific agricultural structures, not global in suitability or scope. Plus, such agricultural systems are only 3,000–4,000 years old; could we have become a planetary force, I wondered, earlier still? As I contemplated this, Wally and Bill's rigorous yet semiquantitative thinking suddenly offered *just* the necessary framework for analyzing the problem. As my focus on the "paleo" possibilities intensified, I discovered the Ruddiman hypothesis,[503–506] and the agrocentric view of recent Earth history it advances, and became instantly hooked.

Time will tell whether the Ruddiman hypothesis eventually prevails over alternatives, but a recent synthesis[573] is promising. It depicts global agriculture and pastoralism all arising and rapidly spreading 8–10 thousand years ago, eventually plateauing in the last 2–3 millennia as our diets gradually shifted from hunting and gathering to preplanned agriculture. These observations mean that even seven millennia ago—when humans numbered some 5–20 million,[380] a mere 0.1–0.4% of our current population,[440] before the onset of widespread human land appropriation[286,319,573]—our diet had already fundamentally shaped Earth. And this dominance was acquired with nothing but rudimentary firewood-fueled technology, when our life expectancy was likely less than half of today's,[604] millennia before the rise of the great rice cultures.[356] Even if the natural mechanisms camp eventually prevails over the Ruddiman hypothesis,[503–506] the above still holds, but "only" for the last 3,000 years or so, some 2,900 years before synthetic fertilizer came to dominate agriculture. Agriculture has thus been reshaping Earth not for decades or centuries, but for millennia, predating the reign of Cleopatra in Egypt, say, or the early rise of Christianity[520] by at least one and quite possibly several millennia.

So, the reality of our diet as a planetary force with Earth-wide impacts is hardly novel, though it may take time to wrap your mind around. The steady rise of methane and CO_2 during the Holocene places these impacts in a broader perspective, offering whole planet "vital signs," earthy counterparts to body temperature or circulating blood cholesterol levels.

At first blush, this pours serious cold water on idyllic "back to nature" farming sentiments[53,170] that strive to heal our broken relation with Earth through simpler, more "natural" ways of eating. If so few of us, eating such simple diets while relying on so little technology, have already altered Earth discernibly, how much "healing" can be realistically expected of inevitably relatively minor tweaks to modern agriculture when there are eight billion of us or more?! We will definitely wrestle with this in upcoming chapters.

More broadly, is it at all surprising that we modify our planetary environment? Despite some US senators' idiotic crowing to the contrary,[80] in this, we are hardly alone.[330,480] Think of beavers raising water levels in ponds and streams, or grazing gazelles selectively foraging certain plants. Not all such impacts are local, either, as Earth's geological record unequivocally indicates. Likely the most spectacular example is a defining event in Earth history, the Great Oxidation Event some 2.5 billion years ago, wrought at least partly by a single celled organism—cyanobacteria—that managed to radically modify all corners of its planetary environment by releasing oxygen—the then widely toxic waste product of newly evolved photosynthesis—into the air. This was later enhanced by a further rise in atmospheric O_2 concentrations by land plants.[339] Prior to these events, for the first roughly two billion years of Earth history, no O_2 gas was present in Earth's atmosphere to speak of, and life was either absent or limited to unicellular organisms. Yet, small though they are, these microbes set the stage for the most recent two billion years, when rising free O_2 gas in the atmosphere both permitted and followed rapidly evolving complex life.[331] And "all corners" it is: from the deepest oceanic abyss, where iron switches behavior once dissolved O_2 becomes abundant in seawater, to continental rocks, whose metals must first "rust" before atmospheric oxygen can build up, to stratospheric heights, where free oxygen absorbs incoming solar ultraviolet radiation and forms ozone, which protects life below from radiation induced biological aberrations.[62,586]

If such "humble" life forms can completely chemically alter Earth's atmosphere and mineral crust, and radically redirect Earth's chemical, physical, and biological evolution, shouldn't we *expect* that eight billion of us or more—each weighing as much as some 10,000 trillion cyanobacteria—have planetary scale impacts?! That is exactly what the late Holocene methane and carbon dioxide records reveal. Neither local nor regional or ephemeral, but planetary and enduring, they tell a story in which feeding ourselves has been fundamentally altering Earth's environments for quite a while, and is continuing to do so ever faster and more broadly.

This book is about these impacts, the planetary imprint of agriculture, rigorously examined through the lens of basic natural science.

I

Preliminaries: Current Food Debates

2

Current Impacts of Diet and Agriculture on Health and the Environment

A reasonable launching spot for our journey into the important planetary roots of agriculture's environmental impacts is the current state of affairs. This mostly involves straightforward science, but it occasionally involves the delicate matter of food choices that we can—individually or collectively—change today for health and environmental betterment, and that constitute a powerful lever we can use to tangibly reduce environmental impacts, to which today's agriculture contributes prominently.

Basic Food-Related Challenges and Opportunities

About 20–30% of all net global greenhouse gas emissions arise from feeding ourselves,[401,450] putting agriculture only slightly behind such notoriously carbon intensive sectors as transportation or industry.

Agriculture accounts for about 70% of all global consumptive fresh water withdrawals.[437] Pasture plus cropland jointly claim 35% of the global continental area and 50% of what is deemed habitable,[437] an area 20–50 times larger than that claimed by all human dwellings. This is important, because agriculture not only occupies vast areas, it often completely dominates and unrecognizably alters them.

Soil erosion, which critically undermines land's ability to provide food, is manyfold faster in agricultural settings, especially croplands,[68,173] than in natural landscapes. Because such erosion rates handily outpace natural soil generation rates,[406] we steadily lose agriculturally essential topsoil at a pace vastly higher than natural soil loss rates.[158]

Agricultural nutrient rich runoff accounts for 65–70% of all domestic discharge of the leading water pollutant, reactive nitrogen.[212,275] (Nitrogen is "reactive" when biologically available for most plants on account of being bonded with hydrogen, carbon, or oxygen rather than with a fellow nitrogen atom, as in the ubiquitous, quasi inert atmospheric nitrogen, N_2.) Agriculture, mostly red meat production,[156] is also a major source of air pollution,[12] exacting an average US death toll almost double that of pandemics.[174]

Agriculture is the key driver of global biodiversity loss,[105,161] which undermines us in many ways, including enhancing the risk of infectious diseases,[44] aggravating already high global inequality,[582] or endangering our water supplies.[185]

Finally, through its strong dependence on climate, agriculture amplifies the direct human toll of climate change, with climate-related dwindling and/or less predictable food production propelling migration and political instability in developing countries.[126,184,306]

These sobering statistics are not really subject to much debate. Yet they apply to the average diet, whereas individual diets range widely in resource needs. This range means that some diets can reduce resource use very significantly while others can raise it considerably, and that two individuals who eat the same amounts of protein can exert radically different environmental burdens.

Comparing Beef to Poultry

The ways we eat thus impact greatly our resource needs, and can affect discernibly, sometimes decisively, national and global resource use. Let's examine some examples, from the body of work my long collaboration with Ron Milo of the Weizmann Institute of Science in Israel has generated. This work has expanded the scope and bolstered the rigor of earlier scientific work,[464] and benefited conceptually greatly from well-researched popular work on the planetary impacts of agriculture.[60,333,469,470]

In 2014, we found that high quality cropland (as opposed to rugged, arid rangelands that can only yield human food if grazed) allocated for

CURRENT IMPACTS OF DIET AND AGRICULTURE ON HEALTH AND THE ENVIRONMENT 13

beef production yields only 15–25% of the protein or calories the land would have yielded if allocated to poultry production, while emitting tenfold more greenhouse gas.[179] In 2016, we thus already knew that raising beef gets you only about a fifth of the desired output, food, while undermining climate tenfold faster. Wishing to extend these results, we wondered[539]: given the relative interchangeability yet very different resource needs of chicken and beef, what would happen if all Americans replaced beef in their diet—at the time about 65 g, delivering 190 kcal (or kilocalories, what food labels casually refer to as "calories") and 9 g protein daily—with the same energy or protein from poultry meat? The answers are reported in figure 3 of our 2016 paper.[539] Because poultry production requires so much less cropland, replacing beef with poultry greatly reduces cropland area needs. We envisioned reallocating that freed cropland to mixed production of items Americans like (as judged by popularity in average American diets). The astonishing answer is that this rather minor dietary shift would deliver enough additional protein—6–9 million kg per day—to sustain *in full* some 120–140 million additional people, or over 40% of the US population, while greatly enhancing the health of those who switch.[42,52,655] The added protein stands to markedly improve global undernourishment, which affects about one in every ten people globally and rising.[403,478] For example, assuming 0.8 billion undernourished people globally, each missing on average about 20 g protein per day, collectively they lack 16 million kg of protein a day, of which the extra 6–9 million additional kg protein per day the beef-to-poultry shift would deliver amount to 40–55%. This minor, readily achievable dietary shift, replacing beef with protein conserving equivalent amount of chicken, implemented in the US alone, can thus realistically halve world hunger.

We can also examine the benefits of the beef-to-poultry shift through the caloric prism. Following the UN definition,[201] let's assume "hunger" means ingesting 1,800 calories per day. Multiplying the difference between 1,800 kcal per person per day and a characteristic personal need of about 2,100 kcal per day by the number of undernourished people[201]—300 kcal per person per day × 0.8 billion hungry people—yields a global daily shortfall of 240 billion kcal. Comparing this shortfall to the caloric benefits of the beef-to-poultry shift—300–350 billion extra kcals per day—shows that replacing beef with poultry can fully supply[299] the global caloric shortfall. If nimble distribution occurred globally (granted, this big if is a key reason for today's hunger), the benefits of this decidedly unambitious transition would therefore eradicate global hunger. It makes sense that two rather distinct definitions of hunger—based on protein or caloric deficit—will yield somewhat different estimates. Yet their message is consistent and clear: every wealthy nation person who replaces their beef with poultry removes 2–3 people from the ranks of the undernourished.

There are fundamental reasons why forgoing steak for something like roast chicken is so impactful. In the same paper,[539] we calculated the overall efficiency of the beef production pathway from producing feed protein by such crop plants as corn or wheat all the way to edible beef protein. The result shocked me: of every 100 kg of feed protein cattle are fed, we only get 3 kg of beef protein; 97% of the protein is lost. Meanwhile, the conversion efficiencies of feed protein into human edible food protein by other livestock types are 10–20%. We then showed that while US beef is relatively carbon efficient by global standards, it is uniquely greenhouse gas–intensive compared to beef production practices in several other developed nations.[461] Demonstrating the robustness of these results, others obtained similar ones in other settings.[117,427,456,475,566,598] We'll get more into the dependence of beef carbon intensity on production methods soon. This will help explain these disparities and, most importantly, why grass fed beef, the most commonly presumed exception to beef's resource intensity, certainly does not improve matters, and in most cases in fact increases beef's contributions to global warming.[94,177,456,461]

Don't let the significance of these results elude you. Since the dawn of civilization, one of our truly existential limits has been good land on which to grow good food. In today's industrialized global agriculture, the significance of land has only increased,[107,529] particularly as our global population grows and stretches Earth's physical limits. For example, since 1960, while the global human population has more than doubled (roughly from 3 to 8 billion[440]), and our standard of living, with which meat eating is positively correlated, has been steadily rising on average, global cropland area has risen by a mere 16%.[200] If there was any more agricultural land to sensibly expand into, we would have claimed it, "westward expansion"–style.[618] This—270% global population increase

CURRENT IMPACTS OF DIET AND AGRICULTURE ON HEALTH AND THE ENVIRONMENT   15

accompanying a 16% increase in cropland availability—vividly illustrates the tightening vise of dwindling arable land. In a very real sense, therefore, cropland means life, and its dearth spells famine and death. In light of this, knowing that 97% of the protein delivered by croplands for producing feed for beef is wasted, or that replacing beef with chicken in the US could fully feed over 100 million additional people, is deeply troubling.

Meat with Plant Alternatives

Even greater improvements would result from replacing all US meat consumption with eating mixed plant alternatives. First, this replacement stands to avert emission of about 325 billion kg of CO_{2eq} annually. (CO_{2eq}, or carbon dioxide equivalent, is a unified measure of the impacts on global mean surface temperatures of emissions of the various greenhouse gases, here CO_2, methane, and nitrous oxide.) It also stands to free up thirty-eight million hectares[183] of cropland (a hectare is 10^4 m^2 or about 2.5 acres), about a quarter of all US cropland. If then reallocated to producing potatoes, peanuts, almonds, or kidney beans, this freed cropland can produce enough additional protein to fully meet the requirements (generously assumed 65 g per person per day[215]) of 0.6–0.9 billion people. Even if repurposed for production of eggs, poultry, or dairy, this spared cropland can still fully satisfy the protein needs of 0.1–0.5 billion additional people. At least to me, such comparisons cast an unusually harsh light on our collective failure to modify our diets.

These results[539] build on earlier ones,[179,180] in which we compared the resource use by the main livestock types and three globally important staple crops. Focusing here on land, we found that a hectare of cropland (as distinct from rangeland) allocated for producing rice, wheat, and potatoes yields about 2–12 times the calories it does when allocated to egg and poultry, pork, or beef. Very similar results also hold per kg protein. Worse yet (for beef), these results completely ignore beef's use of rangeland (defined shortly), which amounted to about 90% of its total land use. Yet even with this questionable treatment of one of the key resources beef use, rangeland,[309,357,534,584] as environmentally free and inconsequential, a cropland hectare still yields 2–10 times more calories when allocated to non-beef food items than it does when allocated to beef.

But Is Using Rangeland Free?

Does discounting the use of rangeland, over which beef lord almost unopposed, make logical, environmental, or societal sense? The typical justification for this discounting is that rangelands can produce no human food other than ruminant meat (which in developed wealthy nations is almost synonymous with beef). Evaluating the merits of this view and answering the above question hinges on appreciating the distinctions between cropland, grassland, and rangeland. In turn, understanding these distinctions requires understanding the nature of different feed types, which we do briefly here, and in detail in the Accompanying PDF.

Distinguishing cropland, grassland and rangeland involves understanding the interplay between precipitation, topography, and cattle feeding. The precipitation impact is intuitive: lush lands produce more grass than similar dry ones, and the same land under the same mean precipitation would produce more grass with regular, reliable rains than with rare deluges briefly punctuating extended droughts. Topography matters because modern cropping relies heavily on machinery, whose use requires smooth, gentle landscapes rangelands are decidedly not. Cattle physiology also affects land classification because rumination allows cattle to extract energy not only from fat, protein, and simple carbohydrates or starch, like we and other animals with a single high acidity stomach (the monogastrics) do, but also from complex carbohydrates. Collectively called "fiber" in nutritional labels, such large carbohydrate molecules mostly survive the monogastric digestive tract, remaining too large to pass the gut–blood barrier, which is critical for productive utilization (reaching individual cells after entering the bloodstream). They are thus indigestible by monogastrics, who reap essentially none of their energy benefits. But rumination ingeniously relegates the digestive task to symbiotic bacteria that live happily anaerobically in the uppermost of cattle's four digestive chambers, the rumen. Unlike feeble humans, these bacteria *can*—by possessing the genetic blueprint for synthesizing the required enzymes—break down complex carbohydrates. By breaking them down into smaller molecules that can easily cross the intestine walls into the blood, rumen bacteria render this stubborn energy source available to the host cattle. Because complex carbohydrates dominate forage, cattle can graze and mix diets like ours—collectively "concentrates" in cattle feed lingo—with forage. Cattle also eat "byproducts," such residues of industrial processing of plant food as citrus peel, sugar beet pulp, or the otherwise discarded residues of brewing corn for ethanol.

This all comes together because rangelands—where rugged topography combines with low, capricious precipitation—typically yield grazed forage of average or lower quality. Lush lands with accommodating topography (such as in eastern Kansas, say, or coastal Texas), conversely, can be used for *either* grazing or growing crops that are used to feed either livestock or humans. Croplands thus offer the most food production volume and flexibility, and on them beef production competes with production of most other human foods.

In terms of human food production, rangelands can only yield ruminant meat, and only inefficiently, because their challenging topography and hydrology reduce their productivity to the lowest of all agricultural lands, far below that of croplands used for grazing. But this focus ignores other uses for these vast, undeveloped lands, notably enhancing wildlife and such ecosystem functioning as reducing soil erosion or dust lofting, or filtering water and promoting stream ecological integrity. Rangelands thus offer either meager beef production or biodiversity and ecosystem integrity.

Considering All Rangeland Uses

So, given their low bang for the buck (less forage mass per hectare per year), does it make societal sense to treat rangelands as a free resource the beef industry gets to use with no environmental accountability because this land cannot deliver any human food other than beef? This question directly pertains to the most common counterargument to the exceptional resource intensity of beef summarized earlier, that beef is desirable nonetheless because it converts sunshine energy captured by grass to a human-destined protein while using little more than rangelands, a resource deemed to have no other value.

Regarding the first element of the above view, does beef really use little but rangelands? In most real (i.e., not imagined) beef operations in the US and other wealthy nations, the answer is a clear no. Let's consult one of the most careful, robust studies of beef's resource intensity, which I'll therefore refer to often, led by Nathan Pelletier.[456] The analysis modeled resource use by a Midwestern beef herd of 100 cows and their associated animals (mostly replacement heifers and growing steers) raised by one of three common practices. Those differ in several ways, the main being the rations on which "finishers"—cattle during the several-month period between weaning at roughly age nine months and slaughter—are fattened to market weight (550–650 kg). In the Pelletier analysis,[456] as in real life, these rations range from what is typical of feedlots, mostly concentrated feed with as little high quality roughage as ruminant digestion permits, to exclusive reliance on forage (so called "grass fed" beef). If the nothing-but-sunshine idea held water, you'd expect the resource needs of the grass finished beef to be lowest among the scenarios studied.

One measure of resource needs is the herd's energy consumption. The Pelletier team found that while in the most intensive practice, feedlot, the herd consumed 714 GJ (714 billion joules) of fossil fuel energy, and the "grass fed" herd consumed 830 GJ while producing less beef. Taking both factors by dividing cumulative energy use by live weight beef produced yields about 38 and 48 MJ of used energy per kg live weight beef produced. So, grass finishing not only does not offer any energy edge but in fact requires 26% *more* energy per unit of beef. The difference— about 90 MJ per kg of beef protein produced—is enough to run a typical microwave oven for a full day. The emission results are consistent, about 15 and 19 kg CO_{2eq} per kg live beef produced. Grass finished beef thus contributes about 30% more than conventional beef to raising atmospheric CO_2 concentrations and surface temperatures.

So much for grass finished beef producing high protein human food out of nothing but sunshine. We will revisit this conclusion for strictly grass fed beef later. Spoiler: it doesn't change markedly, and definitely not qualitatively. Worse (for beef), the herd Pelletier analyzed[456] occupied a lush, topographically accommodating Midwestern *cropland*, definitely not rangeland, further nullifying the nothing-but-resource-we-can't-use enthusiasm for beef. But what of beef raised on extensive ranches in, for example, Nevada or Wyoming, rangelands that genuinely require almost no fossil inputs when grazed? Then the defense partly holds, *when grazed*. But can cattle grazing continue under a foot of snow? In temperate latitudes (including the UK, Europe, montane Asian steppes, and North America, even as far equatorward as central Florida[580]), grass is only available in a 3-6 month long growing season. In the remaining 6–9 months, cattle cannot graze and is therefore served processed forage, hay or silage. Because producing those relies on machinery, they are grown on cropland, and thus compete with crops that could've fed people directly, again at odds with the nothing-but-resource-we-can't-use beef advocacy.

This environmental issue is accompanied by an economic one. To justify these cropland allocations, yields of processed forage, predominantly corn, sorghum, or alfalfa, must be high. This requires high inputs of fertilizer and other agrochemicals whose production is also fossil energy–intensive. The result is that during the 6-9 off-season months, beef feast on fossil energy. In fact, of all the calories US beef eat, no more than a third originates on rangelands, inclusively defined; 65% is derived from fossil energy and cropland-based feed crops.[179,180] Beef's pre-weaning rangeland phase thus indeed features *relatively* low emissions, with two important caveats. First, these emissions are dominated by methane, whose climate impacts a decade or so after being emitted are much higher than that of CO_2. Second, this rangeland phase commits us to a large subsequent expenditure of fossil energy, an elaborate assembly line for sunk costs. We will reexamine this from other angles later, but the image of beef as a food source that uses nothing but sunshine and rangeland is clearly a trivially falsifiable fantasy.

Beyond not offering carbon savings, cattle rangeland grazing interferes with other important societal goals. Large portions of North America and Europe's rangelands are uniquely biodiverse,[285,559] harboring life forms whose survival is a most pressing environmental concern.[151] Perhaps the best known and most controversial example involves wolf reintroduction. Because of its apex predator status in the western US ecosystem, the widespread adverse impacts of wolves' absence from the landscape are well documented, in Yellowstone and beyond.[54,494] It is also unquestionable that the main obstacle to their broader reintroduction involves conflict with livestock grazing,[26] especially the exaggerated concerns about predation,[34,640] or disregard for the ways these predators contribute to landscape integrity.[268,438]

And the harms of cattle grazing run deeper than conflict with wolves. In the aridity of the western US, a key affected landscape is riparian buffer zones, which separate streams from their arid environs and provide an important wildlife habitat.[530] In fact, in the desert Southwest, the primary degrader of riparian ecosystem is grazing livestock, which threatens and endangers wildlife[269] as diverse as the yellow-billed cuckoo, Gila chub, Gila topminnow, desert pupfish, loach minnow, and spikedace. Even the US Department of the Interior's Bureau of Land Management, the notorious agent of destruction[309] of the very lands it is charged with protecting (as is made clear by the Bureau's own website, peer.org/mapping-the-range, and by a July 22, 2021, suit brought by the The Center for Biological Diversity[82]) cannot help but describe the Gila Box Riparian National Conservation Area, a prime example of a riparian ecosystem undermined partly by cattle grazing along the Gila and Sán Francisco Rivers, as[64] a "year-round desert oasis" and "a very special riparian ecosystem abounding with plant and animal diversity." While some promote cattle grazing by invoking the cattle-as-the-new-bison argument, the effects of cattle and bison on ecologically delicate riparian habitats appear entirely distinct; whereas bison help restore damaged riparian habitats, cattle undermine them,[70] so that cattle removal from perturbed riparian buffers powerfully restores them.[38]

Cattle and wildlife are thus in direct competition for rangeland resources,[147,309] with the former almost invariably the victor. Cattle compete over water with naturally occurring species,[148] often displacing them,[614] not via competitive exclusion,[218,245] reflecting cattle's exceptional fitness for these water limited environments, but indeed despite its absence, which human legal and physical protection perversely reverses. On what grounds, then, should we deem resource allocation "useful" or "rational" only when this allocation leads to production of human food? There are clearly many ways land can become valuable and worthy of protection,[465,647] and maintenance of wildlife, biodiversity, or ecological functioning—which, as the above references unequivocally show, western rangelands do disproportionately well—is most certainly chief among them.[191,217] And if you are still unpersuaded, perhaps because you share economists' undiscerning fondness for cost–benefit analyses, consider this: car-deer collisions injure 26,000 and kill 200 people per year, and cost over \$8 billion annually,[270] almost 20% of the value of US cattle production.[612] Or, consider the need to regularly deepen wells to chase receding water tables following beaver eradication.[104,195,554] A small price to pay for the pleasure of eradicating varmints inconvenient to the livestock industry,[106,238] or is it? After all, the beef industry provides about 800,000 jobs,[593] a whopping 0.5% of the total civilian employment (140–160 million). No, leaving some lands alone[287] to sustain wildlife and clean water, and to offer ecological checks and balances, is not only the right thing to do but indeed the *economically* prudent thing to do.

Now that you better appreciate beef's outsized role in the environmental impacts of grazed rangeland, what would it look like to prioritize restoration of the land instead? While at the Radcliffe Institute for Advanced Study, I explored this issue with Alon Shepon—by then a postdoc at Harvard's Chan School of Public Health and now a rising star at Tel Aviv University—and Akshay Swaminathan, a brilliant Harvard sophomore, the only person since Hippocrates to be admitted to every top medical school he applied to (he picked Stanford if you really need to know). Together, we expanded the dietary shift work introduced earlier[183] by envisioning Americans replacing all meat, not just beef, with a nutritionally balanced, diverse plant-based diet that rigorously meets or exceeds the requirements for most major protective nutrients while reducing intake of adverse nutrients. For example,[183] relative to what the replaced meat supplies, these plant-based alternatives enhance delivery of protective carotenes, lutein, folate, sterols, numerous vitamins and minerals, and total and soluble fiber by 50–450%. This is generalizable; the nutrient delivery by plant-based replacement diets are, with very few exceptions (notably the essential B_{12} vitamin), far superior to the animal ones they replace.[52,117,215,397,416]

The environmental boon that this dietary shift would open up is remarkable. To begin with, no beef, no need for rangeland exploitation. This would permit partial or complete rewilding[96,203,648] of about 270 million hectares of US grassland and rangeland, an expanse similar to Argentina, or the combined area of Washington, Oregon, California, and Montana.

Recall that our food production occupies nearly half the global ice-free land area, generates about a quarter of all anthropogenic greenhouse gas emissions, claims almost three quarters of global fresh water consumption, and upsets the balance of nitrogen and other key elements in air and water and on land, greatly expanding the human planetary imprint. Given these impacts, and the observations surveyed above, it would serve us well to question the notion that we are best off using all available rangelands for cattle grazing,[309,471] and subject it to rigorous optimization that considers all conceivable land uses, not just food production. This would identify the grazing levels that best benefit us *all*, and the areas best spared. Among the huge advantages of this commonsensical alternative approach to current rangeland allocation policies is the ability to recognize food-unrelated economic contributions, notably of tourism and recreation, which for many rural communities dwarf those of livestock grazing.[340] This is why it is so important that replacing meat with plant alternatives[183] would save 25% of the US high quality cropland and eliminate 15% and 5% of the national water pollution by reactive nitrogen and CO_{2eq} emissions.

Individual dietary choices can thus really tip the resource use scale. Choosing your diet—especially avoiding meat in general and beef in particular—is one of the most powerful and important environmental decisions you can make. Unlike many other key environmental junctures, this one affords complete freedom from mercurial higher powers, questionable political processes, oblivious corporate executives, timid, ever-triangulating executive branch decisions, or procured legislation by the best Congress money can buy. Forgo beef, and your resource needs drop two- to tenfold. Skip meat, and your dietary impacts on the physical world roughly halve again. Such huge impacts, achieved by fairly simple personal choices over which you have complete control right now, are hard to emulate on any other environmental realm. One integral nationwide measure of this is annual per capita greenhouse gas emissions, 22 and 14 metric tons in their early 1970s zenith and now. This 36% improvement over half a century is dwarfed by the ten- to fifty-fold (1,000% to 50,000%) improvements overnight the forgoing meat shift affords.

3

Of Beef and Public Confusion, Take I: Nutrition

As we have already seen, food choices have had far-reaching planetary impacts, even millennia before industrial agriculture took hold. That's the bad news. But they come with a very good addendum: wise dietary choices can greatly reduce our environmental impacts. We have also began unpacking the validity of various claimed benefits of beef, but—contrary to those—found that replacing beef, any beef, with poultry or plant alternatives confers large and pervasive environmental and nutrient delivery benefits. Confusing.

How Confusion about Beef's Value Developed

Like all reasonably well-read students of food, you know perfectly well that eggs, coffee, soy, meat, beef, carbs, or fat, catastrophically undermine, or magically promote, health. You also know that beef is an environmental apocalypse, or panacea; that plant-based diets will liberate us from, or deepen, our climate troubles; and that such diets extend, or truncate, life and health spans. You read it all, and not in the *National Enquirer*.[302–304,643]

Why is our food discourse, or even more narrowly the nutritional and environmental acceptability of grazing cattle and eating beef, shrouded in so much confusion? A major factor is that, too often, beliefs and anecdotes usurp reliable, robust information, robbing the discussion of the clarity afforded by rigorously quantitative analyses. Let's try to understand this state of affairs, and see what critically analyzed, careful observations can teach us.

We should acknowledge from the outset that nutritional claims do evolve as the underlying science advances. Some of this evolution reflects variability, which is the essence of statistics-based sciences, epidemiology very much included; there is no statistics without uncertainty, because while smokers' lives are shorter on average, some smokers live to 102. So samples must be large enough to faithfully represent the whole population for which they stand, and the small samples that characterize our initial forays into a new research topic are likely to skew our tentative insights. More evolving understanding stems from a yet to be identified need for finer classification of a given dietary element. This explains our changing views of fat once the roughly opposite effects of saturated vs unsaturated fats were recognized, or of carbohydrates once the roughly opposite effects of whole grain vs industrially refined flour became likewise clear. Another source of evolving views reflects the narrow range of our daily consumption, mostly 1,900–2,500 kcal a day. Because of this narrow range, removing one element of the diet almost always implies adding another. Interventions that seek to enact and study one dietary change thus likely induce other, unintended dietary changes whose effects join the effects of the intended change, muddying the observational water.

To illustrate the combined action of these processes, imagine comparing total mortality changes associated with replacing lard, walnut oil, or industrially hydrogenated margarine with white highly refined pancakes or cooked farro. If this were done today, each of the six distinct replacement experiments would have gotten its own dedicated statistical analysis, because we know the experiments are not redundant. But in 1950, we would've considered all as "replacing fat with carbohydrates." Imagine the researchers' surprise when they plotted the results, let's say as the odds of cardiovascular disease diagnoses and proportion of fat in the diet along the vertical and horizontal axes. These plots would have revealed six clouds of largely nonoverlapping points that show that replacing fat with carbohydrates can raise, or reduce, the likelihood of cardiovascular disease, greatly enhancing public confusion and frustration.

But if they then plotted each of the six replacement types in unique colors, six starkly different clouds would have emerged, each color cloud with a unique mean outcome, clearly illuminating the different health impacts of each fat or carbohydrate type.

Further nuance that is confused with uncertainty by some arises from repeatedly revisiting desirability of something from various viewpoints, exemplified by my own work. While much of it has striven to quantify the disproportionately high environmental costs of beef,[175,179,539] in 2021, I explored[175] a complete restructuring of US agriculture that centers on reallocating all lush US croplands to a novel small scale farming system *with beef at its core*, seemingly a 180° reversal. But this apparent about-face is in fact fully explained by the different underlying purpose. While the early papers focused on resource use efficiency, the 2021 paper[175] focused on soil fertility. In it, cattle's main benefit was production of manure of nearly ideal composition for promoting soil nutrient balance, plant fertility, and thus productivity of both feed and food crops, with beef merely a nice-to-have byproduct. If you examine beef's resource utilization efficiency in one analysis, but in another cattle contributions to soil conservation and plant fertility, the incongruent conclusions aren't really surprising, nor unearthing uncertainty in whether beef is environmentally good or bad, because posing such binary questions is childish. It is good, under the very restrictive conditions of the 2021 paper,[175] for soil fertility, and virtually impossible to justify on any other grounds, carbon balance and resource efficiency in particular. But appreciating this and avoiding confusion requires minor homework.

A 2018 paper[182] that explored the possibility and consequences of relying on rangeland-based systems to produce all US beef also caused some confusion. The paper was motivated by the fact that while I am skeptical of and unpersuaded by most of the often-voiced sentiments favoring beef and small scale "local" production (of which Jaster's Op-Ed[284] and Barber's manifesto[35] are nice and long examples, respectively), I remain agnostic as to which approach eventually proves most viable. While this may confuse some readers, and it definitely produced vitriolic comments from vegan activists, who viewed my earlier work as proving their point and the 2021 paper as treasonous, that's what impartial inquiry in the face of uncertainty looks like.

Is nihilism the inevitable response to uncertainty? Does evolving understanding mean that all ideas are equally meritorious? This, we can easily conjecture, is the view Bazarov would have us favor. (Yevgeny Bazarov is a key character in *Fathers and Sons*, Ivan Turgenev's beautiful 1862 novel that popularized philosophical nihilism, whose leading figure, the anarchist Peter Kropotkin, defined as the pursuit of individual freedom and rejection of tyranny, hypocrisy, and artificiality.) No, this is absurd. Nutritional biochemistry or environmental physics are neither tyrannical nor hypocritical or artificial, and most certainly do not wish to curtail individual freedom. In fact, to me, context specific outcomes are scientifically comforting, because they show that all ideas are *not* equally meritorious.

Emotionally, this is similar to our relationships with multifaceted protagonists. Beautifully capturing both this and the key dilemmas *Planetary Eating* addresses is Jim Brewton, a focal protagonist in Conrad Richter's *Sea of Grass*.[488] A retired colonel turned rancher, Brewton is rich, entitled, imperious, often insufferably overbearing, and at times violent. His worst vitriol is reserved for "nesters," homesteaders who favor plowing the native prairie for corn or wheat instead of running cattle on it, which in Brewton's eyes is the only conceivable use for that vast arid expanse. It is *so* tempting to write Brewton off as the detestable epitome of all that was wrong with Westward Expansion. And yet, for students of history, Brewton's contemptuous rejection of "nesting" is prophetic, portraying succinctly a deep understanding of the workings of montane semiarid steppes, and the limits on their use. Recklessly ignoring this knowledge between the 1880s and 1929 is precisely what ushered in the worst environmental catastrophe in US history, the Dust Bowl.[166,654] At heart, *Sea of Grass*[488] delivers a most environmentally enlightened and informed message, but it does so using a most improbable, ill-fitting messenger, as elegant a demonstration of the many gray shades of environmental thinking as I have come across. If we are perfectly comfortable with such ambiguity in novels and films, why should we have any difficulty living with similar indeterminacy in environmental and nutritional realms?

So yes, nutritional science did evolve, and still does. Some half-baked, irreproducible results based on questionable research methods have been published in nutritional science journals, even fancy ones (such as a debate[56] over putative protective impacts of alcohol consumption[47]). Yet far more commonly, scientific shifts and refinements follow sound, careful research whose only fault is that it teaches us something new (hence the "re" in "research"). This is particularly applicable in such sciences as epidemiology that lack an underlying theory embodied in governing equations, the tool that permits physical sciences to make testable, falsifiable predictions. It is also widely applicable when the object of inquiry is as messy and wildly varied in both biology and favored customs as the human body. As distinct from billiard balls on a frictionless table as they come, it is therefore hardly surprising that this murky subject would produce the imperfect, noisy data that epidemiology is saddled with analyzing.

You may then reasonably ask: How do we know that what we now consider "truth" will not be subsequently falsified, toppled like long-discarded earlier nutritional advice? In this, two shifts are on everybody's mind. One is Einstein's demotion of space and time from their Cartesian-Newtonian absolute hegemony. The other is the geography of continents and oceans, which before 1900 seemed immutably fixed, but which Alfred Wegener later showed to in fact be dynamically restless ephemera.[241] But the very reason these two examples are on everybody's mind is their exceptional status as genuine Kuhn's "revolutions."[325] How, then, can we distinguish robust from ephemeral? I have found Naomi Oreskes's[445] semiformal "rules" by far the most helpful in this quest, and have written *Planetary Eating* expressly to conform with them.

In sharp contrast, as I see it, a good deal of public confusion about the nutritional and environmental dimensions of food is intentional. Some arises from deliberate obfuscation,[563] and some from hand-waving extrapolations of tangentially relevant truths or flights of fancy.[255,407]

Let's explore a representative example, a book-length pamphlet titled *Defending Beef* by Nicolette Hahn Niman.[431] You know the book is off to a great start when it singles out *Diet for a Small Planet*[333]—a deeply insightful masterpiece that still towers over the food–health–environment hub—as the source of all misguided, uninformed vilification of beef. Niman then turns to solving the mystery of the ever-rising prevalence of diet-related disease. No, she asserts (p. 2), this rise "cannot reasonably be connected with cattle or attributed directly to butter or beef." Going against the

overwhelmingly dominant view in nutritional science, this pronounce-
ment got my attention. What novel observations enable Niman to finally
solve the riddle? Simple" "because there are fewer cattle on the land today
than there was a century ago."

Are there? USDA data[439] show that between the mid 1800s and 1975,
total cattle inventory has risen roughly fivefold from 29 million to over
130 million, before declining somewhat, hitting 95 million in 2021, 320%
of its mid-nineteenth-century value. So much for fewer cattle. Yet Niman's
argument may still stand. Let's define clearly the line of scrimmage.
Niman wishes to disprove the prevailing notion that beef undermines
health by demonstrating a counter-observation: rates of degenerative diseases steadily rise, while we eat less, not more, beef, the exact opposite of
the prevailing view. Ergo, beef does not undermine health.

The reason Niman's contention is still alive and well despite the inventory error is that cattle inventory is minimally relevant to whether rising
beef consumption explains the rising prevalence of diet-related diseases,
for several reasons. A cow can have a calf every year, or every 3–4 years
when times are lean, introducing 300–400% uncertainty in the relationship between headcount and the variable that is actually relevant to
Niman's argument, beef consumption. Slaughter age can also vary considerably; a steer can take over two years to reach slaughter weight purely on
grass, but as little as nine months on concentrated rations, introducing
another 250–300% uncertainty in the relationship between headcounts
and production volume. Export–import imbalances, a secondary yet ever-
changing factor, decouples headcounts from domestic consumption yet
more. Cattle inventory thus neither corroborates nor refutes Niman's contention, that since we eat less beef over time, not more, beef consumption
cannot explain the rising prevalence of diet-related disease. It is simply
barely relevant to it.

What matters, of course, is how much beef we actually eat. From
the early 1900s through the late 1970s, daily per capita beef availability steadily rose from roughly 50 to 110 g, declined to 65 g in 2015, and
started rising again, reaching 70 g in 2021.[299] And coronary heart disease
prevalence? It rose up until the 1960s, held steady for a bit, and has been
declining ever since.[138,363] Are these trends anti-correlated, as the Niman
assertion requires? No, not at all; while minimally useful, their correlation

is significantly positive, which only holds when the two variables rise and fall in concert, the opposite of the relationship they would have exhibited if Niman's claims were right. But this is all middle school statistics. Health records of 4.5 million patients, methodologically soundly analyzed,[543] reveal the clear truth: increase your unprocessed red meat consumption by 100 g per day, raise your cardiovascular risk by 5–16%. Or, if you prefer, increase your red meat intake from none to 3.5 servings per day, and roughly double your odds of dying.[452] More red meat equals higher, not lower, incidence of disease or death. Now the Niman claim is genuinely, irreversibly dead (Was it a cardiac event?); the more red meat, the likelier cardiovascular disease and death become.

The above observations serve a far more important purpose than merely decisively refuting the Niman assertion, highlighting[172] the role of *confounding variables* in the relationship between beef consumption and disease prevalence. Confounders are attributes of participants (subjects) that we already know impact cardiovascular disease risk. For example, beef-eating smokers who get cardiovascular disease must be deemphasized compared to beef-eating nonsmokers when determining our overall risk of eating beef, because on average, their smoking had raised their cardiovascular risks irrespective of their diet. Additional confounders include manual labor jobs, recreational physical activity,[479] exposure to air pollution, excess weight or obesity, use of highly processed precooked food, among others. Each factor varies widely among subjects, impacting disease risk individually. By amplifying one another's effects, all very likely jointly impact disease incidence even more. The absolute *name of the game* is therefore correcting disease statistics for the impacts of confounding variables, anything that isn't the effect being studied.

Oceans of ink have been spilled about how, given the essential impossibility of controlled experiments on human subjects over the decades it takes for degenerative and metabolic diseases to manifest clinically, we are stuck with the imperfect yet indispensable observational studies as epidemiology's foundational bedrock. And in such studies, apart from wise or at least defensible interpretation, the absolute key is correcting for contemporaneous variability in confounding variables. Lacking any such requisite corrections, Niman's cavalier "data analysis" is without meaning, teaching us nothing.

Same thesis, different venue, several papers[290,670] have recently purported to refute the notion that red meat promotes disease, sadly gaining some popular attention (see, e.g., the September 30, 2019, online WebMD Health News story, "Controversial Studies Say It's OK to Eat Red Meat"). Since their publication, the methodologies[477] and ethics[483,484] of these papers have been questioned. Perhaps most peculiar is their "preferred diet," derived from votes by a 14 member panel, of which 3 were not scientists but "community members," who reviewed people's "values and preferences regarding meat consumption." That is, the "preferred diet" is based on subjective lay persons' personal preferences. Rather than answering "Is eating beef healthy?" the authors instead answered "Do people like beef?" which is both obvious (they do, or else the beef market would have collapsed) and irrelevant (the issue is health, not popularity). Given these failings, the only lessons these papers teach us are not about diet but about the perils of research that is not meant to reduce our epidemiological ignorance but to collect social media likes by "proving" a predetermined point.

What Careful Analyses Show

Let's reorient, turning instead to what actual epidemiological research teaches us, emphasizing larger studies. First is the synthesis mentioned earlier[452] of 3.3 million person-years (i.e., the number of people followed times the duration of their follow-up is 3.3 million). "[A]fter multivariate adjustment for major lifestyle and dietary risk factors" (or "correctly addressing confounding variables"), the synthesis revealed that relative to consuming no red meat, every additional daily red meat serving increased overall death rates by, on average, 13%, arising from 18% and 10% higher cardiovascular disease and cancer death rates. It also found that for every daily red meat serving replaced by such alternatives as fish, poultry, nuts, legumes, low-fat dairy, or whole grains, mortality *dropped* by 7–19%. These cancer results generalize an earlier meta-analysis of multiple prospective studies,[100] which found a 5–31% and 2–33% increased risk of colorectal and colon cancer incidence per 100 g of daily red meat consumption.

Even more extreme results were obtained by meta-analysis of results of eighty-six cross-sectional and ten cohort prospective studies.[153] It revealed significantly lower body mass index, total and LDL-cholesterol, and circulating glucose levels among vegetarians and pure plant eaters relative to omnivores, as well as 18–32% lower incidence of and mortality from ischemic heart disease and 2–13% lower incidence of total cancer. Pure plant eaters fared better still, with total cancer incidence 5–25% lower than that of omnivores.

Another follow-up study of 3.2 million person-years[453] found—after making the mandatory "adjustment for age, body mass index, and other lifestyle and dietary risk factors"—a positive association between red meat consumption and prevalence of type II diabetes; the more red meat, the likelier diabetes becomes, by 12% for every additional daily red meat serving. As we encountered earlier, the same group then extended their inquiry also to various mortality causes.[452] Analyzing about three million person-years (with a roughly 25:75 men:women ratio), again carefully accounting for confounders, they found that the risk of dying from all causes, from cardiovascular disease, and from cancer rose by 7–20%, 13–23%, and 6–14%, respectively, for each additional daily serving of unprocessed red meat (whose disease risks are lower than processed meats[414,651]).

But wait, red meat promoting diabetes? This sounds most counterintuitive, because type II diabetes is about carbohydrate metabolism dysfunction, and red meat contains virtually no carbohydrates. Are the above results real? A recent twenty-four-year follow-up study of over 2.1 million person-years[655] of US health professionals delivers an authoritative yes. It reported the odds of developing type II diabetes three years after replacing red meat with an alternative protein source. Corroborating the Pan et al. 2011 results,[453] type II diabetes risks indeed dropped for each replaced daily red meat serving; by 10–25% when the alternative protein source was poultry, 2–23% with seafood, 14–22% with low-fat dairy, 14–23% with high-fat dairy, 1–19% with eggs, 2–18% with legumes, and 11–22% with nuts. The authors address these varying results, suggesting that "intake of red meat and the other investigated protein sources might be associated with different dietary patterns and lifestyles" and that "benefits of these substitutions are likely due to multiple simultaneous changes in nutrient intakes, e.g., fatty acids, proteins, heme iron, sodium, dietary fiber, minerals, phytochemicals, and other bioactive components."

It is technically possible that all or most of the above benefits of replacing red meat may thus be attributable to the dietary pattern that accompanies meat, not the meat itself. While improbable, and at odds with results that address meat itself or that correct for this effect, this notion cannot be decisively rejected. Yet the real problem with this argument is that given the powerful impacts on public health statistics of these associations, they must be very ubiquitous and recalcitrant. While it is not impossible to overcome them (i.e., eat only the red meat with none of the accompanying white bun or sugar and salt laden condiments that these associated patterns comprise), every eater should therefore ask themselves how likely it is that they will be the rare exception who successfully bucks these pervasive, entrenched associations and avoids the dietary patterns with which red meat consumption is so intimately associated. I am skeptical of this possibility, because exceptions are, by definition, exceptional. If it were any easier to avoid the associations, enough people would have taken advantage of this possibility, reducing the strength of the association.

Consistent with the above work, plant-based diets have an edge in most studies. Combining and meta-analyzing twelve prospective cohort studies with about half a million participants revealed[279] 1–18% and 14–30% reduction in all-cause and coronary heart disease mortality among the group with highest adherence to a pure plant-based diet relative to the lowest adherence. This led the authors to ascribe a "potential protective role against chronic disease mortality" to plant-based diets.

Using the same approach, Chinese and Harvard researchers[628] corroborated the above results, but with a twist. Among US participants, they found a 12–19% increase in all-cause mortality per each additional daily serving of unprocessed red meat, mostly due to elevated cardiovascular and cancer risks. But the team found no such associations in the international population, suggesting again that something beyond beef is a factor that lurks behind those associations, that it is not beef alone that is deleterious, but a cluster of lifestyle patterns that—at least in the US—are associated with elevated beef consumption.

Consistent but definitionally distinct associations were also found in several other prospective studies of US healthcare professional cohorts.[397] Analyzing about 3.5 million person-years, Song et al.[397] found—after the usual adjustments—that risk of cardiovascular disease related death rose by 1–16% for every 10% increase in animal protein intake (notice the broader definition), but declined by 5–14% for every 10% increase in plant protein intake. In another interesting twist, these associations disappeared among individuals with no known risk factors; neither all-cause nor cardiovascular disease death risk was significantly associated with consumption of either plant or animal protein.[397] These disconnects also raise the possibility of as-yet-unknown factors that accompany elevated beef consumption. However, the ubiquity and reproducibility of the strong population-level link between red meat consumption and various measures of mortality[188,207,254,669] again suggests that even if choices that tend to accompany beef eating, not beef eating itself, are to blame, the association is exceedingly difficult to escape. Most of us are thus unlikely to dodge the beef–disease association. Additionally, given that age is a key risk factor for most diseases, and that lifespans are increasing, fewer of us would fall into the "no risk" category over time.

A Better Nutritional Route

Why, despite the above evidence, do some continue to argue that meat, particularly beef, offers a superior nutritional package that no plant combination can match, and that, therefore, "beef is good for you"? One nutrient that often comes up often emphasized in this context is conjugated linoleic fatty acid (CLA), a naturally occurring trans fat abundant in beef and milk. While industrially produced trans fats strongly promote cardiovascular and other diseases,[76,142] CLAs are protective. Laboratory studies suggest that CLAs reduce cancer cell viability,[322] and animal models have shown CLAs to effectively combat atherosclerosis, diabetes, and inflammation, and to promote immune response.[389,631] This anomaly is especially pertinent to the conventional vs grass-fed beef debate we'll get into soon, because the CLA content of milk and beef is higher in grass fed than in grain fed animals.[433,617] This and similar, narrowly focused results are then generalized by some, who also invoke elevated risk to plant eaters of reduced bone mineral density or insufficient intake of micronutrients or protein.[31,274]

But plant eaters can easily avoid these risks, notably the overblown protein concerns,[368] as lucidly summarized by Katz and Gardner.[215,303,366] In fact, both scientific[495,649] and anecdotal/personal[293,487] evidence shows that adequate protein intake on a strictly plant-based diet is distinctly realizable even among endurance athletes whose protein requirements for muscle repair greatly exceed the average person's.

In addition, our work has shown that thousands of different plant-based diets that supply desirable levels of over twenty health promoting nutrients exist and are fairly easy to obtain.[181,183] There is thus more than ample evidence that purely plant-based diets can provide enough varied and diverse protein[537] and promote health[388,514] to lay to rest any micro- or macronutrient intake concerns about subsisting on plants alone.

At the same time, some comparisons do challenge the above results, demonstrating associations of beef eating with equal or even superior health outcomes. For example, replacing beef with refined white flour pasta is likely to undermine health.[239,262,553] Likewise, vegans who are careless about B_{12} intake can become deficient in this essential vitamin, which only animal products or supplements provide. Like the Wang et al.[628] results noted above, another striking result arose from analyzing some 0.8 million person-years in Japan.[419] Comparing in a methodologically and ethically sound[255,407,502] manner those who consumed the most and least meat (spanning an almost eightfold consumption range), the authors found *no* difference in death rates due to ischemic heart disease, stroke, or total cardiovascular disease. They thus concluded that consuming 100 g meat per day or less does not raise death rates from those circulatory diseases. The nutritional meaning of such studies has yet to be decoded. They may reveal, for example, the need for epidemiological studies to disentangle the association patterns discussed above, distinguishing the health effects of protein derived from solid muscle versus ground beef, or of hamburger mix per se from the hyper-processed profusion of salt, sugar, and preservatives that usually accompanies burgers in fast food chains. For now, such studies constitute a distinct minority exception that may require fine-print qualifications but that leaves the rule—that beef is damaging for the modern human body—fundamentally intact. In science, solid majority views, sustained over decades, appear[444] to rarely form around the wrong side of an argument. Instead, they tend to form around ideas that explain the bulk of what is robustly observed most succinctly and simply, with the fewest questionable conjectures, and to eventually be reconciled with contradictory observations by finer resolution (e.g., classifying fats by circulatory effects, distinguishing industrially produced from naturally occurring trans fats, or classifying carbohydrates by impacts on insulin sensitivity).

Another reason these exceptions do not currently call for paradigm shift is provided by intervention studies that document many, varied, and significant health benefits of transitioning to plant-based diets. For example, a compilation of twenty-seven such studies lasting up to two years found[386] "robust evidence for short- to moderate-term beneficial effects of plant-based diets versus conventional diets on weight status, energy metabolism and systemic inflammation in healthy participants, obese, and type II diabetes patients." If eating animal products were important to health, their complete elimination from the diet should have undermined, not improved health.

The "We Evolved with Beef" Perspective

It may be reasonable now to consider a perennially popular argument against shunning meat, that humans evolved as meat eaters, that our ape and hominin ancestors were frequent, even obligatory meat eaters,[86] and that we thus remain obligatory meat eaters. Most recently, this idea was invoked to promote yet another magical diet that would—if only you purchase our $399.99 a day meal plan—reverse weight gain and keep degenerative diseases at bay, the co called "paleo" diet.[130]

The argument mostly focuses on a period in our evolutionary history during which our brain expanded rapidly while our digestive system shrunk, which is taken to reflect a dietary transition from plants to meat. From this, proponents suggest a causal link in which meat was the indispensable fuel that propelled the development of bigger brains during our evolutionary transition from apes to abstract analysts.[4,646,680] The argument may be true, but if the composition of current or prehistoric diets is not particularly well known,[246,364] certainly not in geographical details,[526] how well can we expect to know diets two million year ago?

Moreover, while eating meat in general requires smaller digestive tracts, the validity of this association is limited. At roughly 400–600 kcal and 10–20 g protein per 100 g, nuts and seeds are low volume, high nutritional density foods for which small stomachs suffice. Top them off with peanuts and some honey,[367] and you can do well as a plump, small-stomach, obligatory plant eater.

Want more numerical specificity? Suppose we explore replacing modern beef with various modern plant items, and calculate the masses of those we would need to fully supply all the calories and protein in the forgone beef. Of the two answers we get—one for exactly replacing the energy, and another for exactly replacing the protein in the forgone beef—let's take the more exacting one, the maximum of the two masses. Out of fifty-nine plant items I explored, I found seven (12%)—almonds, kidney beans, peanuts, pistachios, chickpeas, lentils, and soy—whose required replacement masses were in fact *smaller* than that of the beef they replace, 800 g per kg beef on average. Six more—barley, hazelnuts, oats, walnuts, buckwheat, and spelt—required only slightly higher masses—20% higher on average—than the beef mass they replace for the exact full energy–protein replacement.

While Pleistocene forebearers of these nuts and legumes differed markedly from their modern counterparts, the message is clear: if more than two in ten plant items are just as energy and protein dense as game meat, early plant-eating hominins could have invested relatively modest efforts in gathering plant-based diets with no less protein and energy, and no more bulk than large game eating would have provided. This is why I find this argument unpersuasive.

Even if our exceptional analytical skills did require meat to fuel this evolutionary upgrade, given how vastly biologically, socially, environmentally, physiologically, and nutritionally different our lives are from those of our early evolutionary forebearers, how useful is this association to guide dietary choices today? The physiological challenges modern diets must address and biochemically protect us from—long lives with virtually no existentially required activity and unlimited food supplies—clearly have almost zero overlap with the challenges our ancestors faced—predation by larger predators, unpredictable food supply, and brief, violent lives.

But suppose, despite these game changing differences, a "paleo" diet is nutritionally wise for modern humans, is it deployable? I'd say not even minimally, because it is practically impossible to ever find meat, cereals, or greens that even vaguely resemble their paleolithic predecessors. For example, even lean grass fed beef or bison, the nearest crude modern analog to hunted Pleistocene fauna, are still 2–3 times fattier than wild meat[596] and surely even more distinct in micronutrients.[457] Likewise, how similar to their naturally occurring counterparts are for example, manicured arugula or hyper-bred strawberries? Likely not particularly. In my decidedly unmanicured yard, the wild strawberries are as distinct—in shape, size, taste, and abundance—from their grotesquely enormous estranged modern relatives as a modern confined dairy cow is from her Holocene aurochs progenitors. One proxy for this comparison is the difference between organic produce and conventional counterparts, where large micronutrient differences are observed[3,75,272,423] and likely understate the differences we are after, because modern organic produce is anything but wild. While further research is needed, the case for a "paleo" diet that can be reasonably characterized as promoting health in the twenty-first century is yet to be made.[192,364,424] A recent effort[660] to evaluate the paleo diet concluded that the current evidence is far too minimal to recommend it even for the clinical objective the diet is most likely to achieve: diabetes management. An earlier one[90] described the paleo diet as "an expensive and not nutritionally adequate diet with a high carbon footprint." As to the success odds of future efforts to corroborate the putative benefits of the paleo diet, I find it interesting that one effort to examine this topic[288] was published in the journal *Utopian Studies*...

While the nutritional case for "paleo" diet is being litigated in the scientific literature, the environmental case against "paleo" diets is strong, mostly because of its strong emphasis on meat, especially beef, and other animal products.[90]

The above discussion of "paleo" diets is related to but logically distinct from promotion of wild animal consumption by some in the "meat is essential" camp, most compellingly and eloquently relayed by Steve Rinella.[492,493] While interesting, this thinking is irrelevant to the question at hand, because we are interested in population statistics, not one outdoorsman's awesome skills (which are *very* dear to my heart, and which

Rinella beautifully and convincingly showcases). If you are unsure what distinction I am drawing here or why it matters, try recalling the last time you killed your meat, field-butchered it, carried it on your back many miles, and cooked it over open fire, as Rinella describes. However endearing and compelling these exploits are, they are no more relevant to population-level modern diets than Robin Hood is to modern fiscal policies.

The Take-Home

I hope you now understand why, in the face of voluminous, expanding literature about the ill health effects of red meat, clinging to the idea that it contributes to good health seems to me ninety-nine parts theology and one part science. What best characterizes the writings of self-appointed beef "defenders" such as Niman is not facts but an inability or unwillingness to distinguish sometimes nonsensical, and at other times interesting and curious, chaff from the wheat that is the bulk of the relevant literature. Such estrangement from fact is equally prominent in musings that address the environmental dimensions of diet, to which we will get soon. To cultivate a "beef maven" public persona [429,430,632] or to peddle fawning reports on charity-for-the-0.01% initiatives [657] on such foundations is to misunderstand the role of expertise in public discourse [355,445] and brazenly abuse the critical role of actual experts in empowering lay persons to successfully distinguish opinions, logically defensible conjectures, and mature scientific views. [236,304] For those of us who are dedicated to numerically analyzing nutritional or environmental claims about diet, and who take the time to transcend anecdotes and gratuitous assertions and develop the observational, conceptual, mathematical, and software frameworks needed for analyses of the requisite rigor, the description of *Defending Beef* [431] as offering [632] "in-depth expert analyses" is hilarious. But the high stakes, the considerable damage to human and planetary health such nonsense causes, add a distinct sour aftertaste to this hilarity.

I do not mean to single out Niman; she is but one example in a burgeoning class of self-appointed "experts" on food, nutrition, and the environment. And I concede that—being fundamentally premised on doubt—science has limitations, and that however robust the governing science may be, it alone rarely definitively settles all divergent opinions. However, scientific inquiry still remains and hopefully will forever remain the foundational bedrock of knowledge. That is why, as evidence of beef's nutritional and environmental harm has transitioned from novelty[178,464] to mainstream,[182,229,475] most scientists have concluded that beef is irredeemably resource-inefficient[642] and a nutritionally poor choice.[52,239,302,452]

Yet, widely accepted though these conclusions are, they have not gone unopposed. The motivations of the objectors vary from well-intentioned convictions their authors confuse with facts (beautifully illustrated by Bloom's[65] cogent and unpretentious Letter to the Editor), to ostentatiously brazen[167] or ethically questionable[412] protection of backers' interests at the expense of the public's. In the next chapter, we examine several arguments advanced to counter the robust environmental case against beef, with some emphasis on a key recurring counterargument that revolves around cattle grazing.

4

Of Beef and Public Confusion, Take II: The Environment

Let's now turn our attention to environmental dimensions of the grand question: *Is eating meat, beef in particular, a good idea?* The focus on beef will become clear shortly, but the short version is that it is the most resource-intensive, by a comfortable margin, compared to most any reasonable alternative. Analyzing beef is thus an obvious anchor to any serious discussion on the environmental dimensions of dietary choices. Let's kick off our discussion by expanding on chapter 2 regarding the key issues around beef. Because most food emissions arise in the production phase[133] (as opposed in retail or the home), that's where we'll mostly focus.

Beef's Comparative Footprint

Analyzing US beef data from the first decade of the twenty-first century,[179,180] we found that producing a kg of US beef protein (contained in roughly 4 kg or 9 lb of beef) requires about 3,300 m^2-y (1 m^2-y is one square meter, allocated for one year) of total land. These needs are importantly split among cropland (roughly 10%) and rangelands (the remainder 90% or so). Croplands are rich, lush, topographically accommodating lands with ample water (as irrigation or sufficient precipitation) that can produce most foods, not just beef. At least in this book (some may differ), rangelands are, on the other hand, the opposite: meager, arid or semi-arid, topographically challenging marginal lands that can only yield human food if used for ruminant grazing.

Producing a kilogram of US beef protein also requires allocation of 3 kg of reactive nitrogen fertilizer, and emission of 200 kg of CO_{2eq} (defined in chapter 2). The basic dilemma, call it the fundamental environmental liability of beef, is that if used for producing other foods, the same resources would yield much more human food, containing much more protein.

Take, for example, almonds. A kilogram of almond protein requires allocation of 26 m^2-y of cropland, 0.4 kg of nitrogenous fertilizer, and 2.5 kg CO_{2eq} emissions; so we can get about 8, 12, and 90 times more protein from almonds than from beef per kg fertilizer, cropland hectare, and kg CO_{2eq} emitted. Since all resources are needed for either almonds or beef, let's focus on the "limiting" resource, the one that yields the *least* alternative protein, here fertilizer. With 3 kg of nitrogenous fertilizer, we can thus produce roughly 1 kg of beef protein or 8 kg of almond protein. The "8 kg almond protein" option thus uses the same fertilizer mass the "1 kg beef protein" requires, but saves on some of the other, nonlimiting resources.

The cropland savings are beef's needs, 310 m^2-year (approximately 10% of the 3,300 m^2-y total land needs mentioned earlier) minus 8 kg almond protein × 26 m^2-year of cropland per kg almond protein, or about 100 m^2-year. The emissions savings are 200 kg CO_{2eq} minus about 20 kg CO_{2eq} (8 kg × 2.5 kg CO_{2eq} per kg), or approximately 180 kg CO_{2eq}. Favoring almonds over beef thus produces eight times the protein while saving about 100 m^2-year (about half a singles tennis court) and 180 kg CO_{2eq}.

Not all replacements of beef by plant alternatives offer only benefits, as less protein-dense apples demonstrate. Producing a kilogram of apple protein requires 225 m^2-year of cropland, 1,210 g of nitrogenous fertilizer, and 275 kg CO_{2eq}. In terms of cropland and fertilizer, apples again have an edge over beef (requiring 225 instead of 310 m^2-year of cropland and 1.2 instead of 3 kg nitrogenous fertilizer). But a kg of apple protein requires the emissions of 275 kg CO_{2eq} instead of beef's 245 kg. With about 15% *more* emissions per kg protein, instead of saving emissions, swapping beef with apples as a protein source actually exacts carbon costs.

Such exceptions notwithstanding, repeating this calculation for sixty-six varied and popular food items[181,183] reveals that forty-five items, almost three quarters, provide more protein than beef (on average about four times more) per unit of the limiting resource. While most of these items are plant based, they also include dairy, eggs, chicken, and pork, which provide 3–9 times as much protein as beef per unit of the limiting resource while saving 8–28% of other resources.

Numerous alternative plant- and animal-based foods can thus deliver on average fourfold more protein per unit resource than beef. Individual items range from one- to twofold for various fresh fruit, to three- to fourfold (broccoli or cabbage), seven- to eightfold (garlic, oats, or barley), ten- to twelvefold (kidney beans or tofu), eighteenfold (peanuts), and thirtyfold for soy. Dairy, eggs, chicken, and pork span three- to ninefold. In favoring these alternatives over beef, you also simultaneously markedly drop your use of the other (nonlimiting) resources, as almonds illustrate above. For some item/resource combinations (e.g., emissions for almonds, walnuts, pistachios, or hazelnuts), these savings exceed 90%. Averaged over all item/resource combinations, they exceed 25%. Combining the added protein produced and the reduced use of nonlimiting resources, our finite resources can yield tens of times more protein per unit resource when allocated to many different beef alternatives than when allocated to beef.

Very few environmental choices offer such enormous savings. Take, for example, gas mileage. The worst gas guzzlers get as little as ten to twelve miles per gallon, while the latest electric vehicle gets—nominally—some 100–120 miles per gallon. So the most celebrated improvement in personal transportation offers only a tenfold improvement, compared with the twenty- to thirtyfold improvement replacing beef with soy or peanuts offers. Or, consider home appliance efficiency, which rose 50–220% over 1981–2013 (according to M. J. Perry's www.aei.org/carpe-diem January 6, 2015 post). That's roughly an efficiency doubling over thirty-three years, versus beef alternatives that offer savings 10-25 times greater immediately. These uniquely high resource needs of beef production explain why the discussion on the environmental burdens of diet centers so firmly on beef.

Examining Pro-Beef Counterpoints

As noted, the conclusions about beef's resource intensity have been vigorously contested, on several grounds, despite the weight and rigor of the observations on which they are based. One objection is that the calculations in which beef is replaced with one alternative food item are unrealistic and unrevealing, because people don't eat one-item diets. Sounds reasonable, but recall that these alternatives replace only the small beef portion in the diet, so people in fact do eat such small portions of their diet from a single item. Even the two slices of bread in a typical sandwich, with nothing in between, hold as many calories as the replaced beef's roughly 200 kcals. An even more powerful rejoinder to this objection is offered by our calculations[181,183] in which beef or, separately, all meat items are replaced by mixtures of many replacement items whose compositions are not random but instead selected to satisfy tens of nutritional inequality constraints (e.g., the vitamin C, protein, or calcium delivery by the mixed replacement diets cannot fall below the delivery by the replaced beef, which is readily achieved using a simple optimization technique called "linear programming"). Despite this rigor, we easily found[181,183] thousands of such mixed replacement diets that offered not only environmental benefits but also nutrient packages handily superior to the replaced meat. This includes far higher intake of various vitamins and minerals, total and soluble fiber, and various sterols, all known to promote health.[405,476,620] The one exception, vitamin B_{12} insufficiency,[181,183] can be cheaply and fully addressed with supplementation[681] or using nutritional yeast. Compared with the meat they replace, these replacement diets are thus likely to improve not only the state of our environment but also public health. These serious environmental and nutritional strikes against beef make defending or promoting it that much harder.

Some counter beef's characterization as resource intensive by pointing to a steadily rising efficiency of beef production.[93] The argument is factually sound but logically uncompelling. First, our and others' estimates[179,475,642] of the added environmental costs of beef are based on analyzing current, not historic, beef. The four- to sevenfold performance disadvantages of beef we calculated[179-183,539] thus already reflect modern, improved-efficiency[93] beef. More important is considering the underlying logic of this point. We do not compare new cars' gas mileage to that of the Model T,[48] but to competing models of roughly the same vintage. We do not reject proposed emission standards because current air quality is already better than that of Dickensian London. We do not reject novel medical intervention because we already outperform village healers, and so on. The same standards should also apply to carefully done studies of food and agriculture.

Another attempted rebuttal of beef's environmental liabilities invokes the observation that some of what cattle eat cannot be used to feed humans. Common examples include grass and industrial byproducts (such as citrus peel from juice pressing, or brewers' grains from ethanol production). Grass and byproducts account for roughly one-third and one-tenth of US beef rations, respectively.[179,180,182] Over 60% of all beef feed thus still requires the use of high quality cropland, fertilizer, water, and greenhouse gas emissions, all of which can be used to produce most any food. The notion that beef and human food production do not compete for resources is thus 60% false. The key exception involves byproducts that, if not fed to cattle, would require financially and environmentally costly disposal. Two weaknesses make this argument unpersuasive *even* for the remaining 40%. One is that there just isn't that much byproduct feed to go around, under 15% of the feed calories in beef rations even at the height of the ethanol frenzy, when brewers' grain supply peaked.[182] The other stems from the fact that byproducts are all concentrated feed types that, given cattle physiology, must be accompanied by roughage (forage), whose production requires expending fossil fuels (see the Accompanying PDF for further details). The bottom line is clear and inescapable: beef do not eat only what humans cannot; they eat that, plus much more that directly competes with human food production.

But if insufficient roughage is the limitation, can we combine concentrated feed from byproducts with roughage from grazing to optimally overcome the limitations of both and use byproducts productively after all? Exploring this possibility,[182] we showed that if deployed with even scant attention to biodiversity,[36,459] this combination would only supply a quarter to at most a third of current consumption.[182,252] The "beef eat what we cannot" argument can thus be made to hold some truth, but only in a broader context of a completely redesigned alternative food system with two key attributes. First, the scope of rangeland use cannot be today's use minus cosmetic reductions, but exactly the complement of byproduct availability, providing only the roughage necessary to turn existing supplies of byproducts into a well-rounded national cattle diet comprising roughly 40–60%[213] roughage by dry mass, and *no more*. Because cattle feed rations based on this allocation model can deliver only 25–35% of today's beef consumption, [182,252] the second key element of this alternative system has to be an ironclad commitment to reducing beef consumption levels to that amount, thereby demoting beef into a calorically minor condiment. Without this commitment, promoting eating beef because cattle eat stuff we cannot is no more persuasive than a one-faced Janus. It also seems improbable that the relatively optimistic grass productivity estimates the above is based on will persist, because of grassland degradation[420,481] by cattle grazing and the marked aridity climate change is inflicting on the already arid western US.[125,365,449,548,644,645] And the lower grass output these constraints predict means an even sharper required beef consumption reduction.

The Most Frequently Invoked Argument Promoting Cattle Grazing

We next turn to the most ubiquitously invoked counterargument to beef's outsized environmental liabilities, the carbon sequestration potential of grazed lands.[136,329,499] This argument is potentially powerful because it has a mechanistically plausible core. (While robust statistics can be powerful, they are much more so if accompanied by simple, plausible mechanistic explanations that not only reveal relationships between various explanatory and explained variables but also explain why and how these relationships hold.)

The first step in assessing this counterargument is to appreciate the distinction between radiatively active and inactive forms of carbon. Atmospheric CO_2 gas is "radiatively active" in the sense that it acts as a greenhouse gas—that is, absorbing some of the infrared radiation the surface emits upward—with the ensuing atmospheric warming leading to enhanced downward radiation from various heights in the atmosphere toward the surface. This slows surface radiative cooling to space, enhancing surface energy retention, and temperatures. In contrast, solid or liquid belowground carbon (notably soil organic carbon) is "radiatively inactive" because by virtue of not being a gas and being stored (sequestered) below the surface, it has no direct impact on surface cooling. (But various proposed or demonstrated pathways for enhancing belowground carbon sequestration can have a huge indirect surface cooling effect, because, if robustly demonstrated to exist and endure, they speed up conversion of radiatively active atmospheric CO_2 into belowground carbon that no longer warms the surface.)

Now we get to relevance to the carbon intensity of beef. If cattle grazing can be shown to reproducibly enhance rangeland carbon sequestration, and if this enhancement is large enough to offset direct emissions incurred during beef production, grass fed beef becomes a useful tool for combating climate change.

While this notion is mechanistically possible, even plausible, does reality cooperate, or do the above proposed mechanisms mostly represent unrealized potential? Does grazing enhance atmosphere-to-soil carbon fluxes? Some biochemical or genetic observations support this proposed enhancement,[500,676] while others are at odds with it, as we will see in detail shortly. I think it's fair to say that we are still unsure about the scope and magnitude of these putative enhancements. But based on current observations, my money is firmly on their proving to strongly depend on geography, climate, and weather. For example, a recent analysis of global dry rangelands[214,359] reveals that as grazing pressure rises, dryland soil carbon storage is reduced in warmer conditions, but enhanced in cooler conditions. For US grazing, this means, for example, that grazing is likelier to undermine soil carbon storage in Nevada or Arizona, say, but to enhance it in the Dakotas, Montana, or Wyoming. In northeastern China steppes, the picture appears simpler[630]: the more grazing, the less soil organic carbon throughout. A recent global meta-analysis[486] that combined numerous published papers to examine the response of soil organic carbon stocks to myriad conditions (such as temperature, precipitation, soil depth, or grazing intensity, i.e., how many animals occupy a given land area on average) corroborates this. From 350 earlier analyses, the authors found a hint that perhaps light grazing raises soil carbon very minimally (by 1–2%) relative to no grazing, but this effect was not statistically significant. Analyzing then about 1,200 additional earlier comparisons of soil carbon stocks under moderate or heavy grazing to no grazing showed, this time significantly, an opposite effect: the heavier the grazing on it, the less carbon the soil contains. Considering the overall effect of grazing relative to no grazing in all environments and intensities revealed that introducing grazers resulted in about a 10% reduction in soil carbon content. Grazing reduces soil organic carbon, not enhances it.

Let's update these results by summarizing five recent meta-analyses, the above plus 4, that combine well over a thousand individual comparisons of soil organic carbon under grazing to that without grazing in widely diverse settings spanning most relevant environmental conditions. Because of the dependence on grazing intensity reported above, let's focus only on light grazing, the grazing intensity that appears to affect soil organic carbon most favorably. The results are reported as the percentage

$$p = 100 \, \frac{\text{soil organic carbon under light grazing}}{\text{soil organic carbon under no grazing}}.$$

With this representation, $p = 100\%$ means that soil organic carbon of ungrazed plots remains unchanged once light grazing is permitted on those plots; $p > 100\%$ means that such introduction raises soil organic carbon; and $p < 100\%$ means that soil organic carbon of ungrazed plots drops once light grazing is introduced. The result summary follows.

| lead author last name | year | ref. | $100 \, SOC_{\text{light grazing}}/SOC_{\text{no grazing}}$ | | |
| | | | most likely ratio | likely bounds | |
				lower	upper
Beillouin	2022	45	98	96	101
Ren	2024	486	102	98	105
Hao	2024	244	98	94	102
Li	2024	345	99	93	105
Zhan	2023[a]	672	91	103	79
Zhan	2023[b]	672	97	100	93

[a]humid sites   [b]arid sites

Note that only the last row presents results (by Zhan et al. 2023, for arid sites) that differ significantly from zero; because the range spanned by the percentage bounds of all other rows contain 100% (e.g., in the first row, 100% sits between 96% and 101%), their least conjectural message is that light grazing does not affect soil organic carbon. That is, if light grazing is introduced on currently ungrazed plots, soil organic carbon sometimes rise minutely, sometimes drops a tad, with no clear effect; light grazing affects soil carbon storage minimally and insignificantly. In other words, the notion that even light grazing—the grazing intensity to which soil carbon appears to respond most favorably—enhances soil carbon storage is no more likely than the competing opposite notion, that light grazing reduces soil carbon storage. Regarding moderate or heavy grazing, the effect is clear: the more of those, the less organic carbon in the soil. The only significantly different from "no effect" meta results is thus delivered by the last row: in arid sites, the addition of light grazing significantly reduces soil carbon storage. Again, reduces, not enhances.

If grazing reduces soil carbon content, it cannot possibly promote sequestration any more than regularly outspending your income promotes your savings.

The current imperfect observational picture thus depicts grazing effects on soil organic carbon as widely varied in time (e.g., following weather variability on two- to ten-day timescales, the march of the seasons, and year-to-year and longer timescales climate variability) and space (e.g., to differ widely among arid, semi arid, temperate, or Arctic locales). Soil organic carbon enhancement by grazing is the distinct exception to the rule that domestic grazing livestock undermines carbon sequestration.

Before systematically examining this, it's worth noting that grass fed beef is likely practically unviable. In wealthy nation markets, strong economic forces discourage grass-based beef production and reward the industrial–conventional path. These forces partly explain why Europe's beef, traditionally exclusively or mostly grass-based, is steadily drifting toward intensive, concentrated feed-based American-style beef,[5,249] and why grass fed beef stubbornly remains a minuscule niche market[40]. These whimsical economic currents are likely a pertinent cautionary tale but are outside the scope of this book. With this out of the way, we are ready to examine further the key question: Can grazing enhance rangeland carbon storage enough to remove appreciable CO_2 mass from the atmosphere to matter to Earth's climate?

For the answer to be a solid yes, improved rangeland management must enhance current atmosphere-to-soil carbon uptake rates significantly and reproducibly under a wide range of conditions. A single positive result, or a set of positive results in uniquely favorable environments, is insufficient. The enhancement—the rate of soil uptake of atmospheric carbon under cattle grazing minus the background rates without cattle grazing—must be large enough to remove significant carbon from the atmosphere. Otherwise, the process would be too slow to matter, like emptying an olympic size swimming pool (the atmosphere) using a straw. In addition, the soil's capacity to contain the added carbon must be much larger than the augmented sequestration, or else the soil will quickly saturate with carbon, precluding further uptake. For our purposes, this criterion is met if rangeland soils can store a large carbon mass, so that the cattle augmented uptake can persist for a long time before the soil saturates.

The reservoir size (latter) criterion appears to hold. Global rangelands[338,575] used for cattle grazing store roughly 300–550 billion tons of carbon, one-to two-thirds of the atmospheric carbon content, 900 billion. (Note that here, a ton—sometimes spelled a tonne—means only 1,000 kg.) So, global grassland soils store enough organic carbon for the sequestration argument to be taken seriously.

Best estimates of atmosphere–soil carbon fluxes over grazed lands fall in the 0.4-0.7 billion ton per year range.[101,327,338,575] Achievable global soil uptake *enhancement* attributable to improved management[39,87,258] can be as low as 0.1 billion tons of carbon a year (ignoring very low values, to cast a more positive light on grazing), yet may hypothetically exceed 0.7 billion tons of carbon a year if all global rangelands are optimally managed.

Can improved grazing management enhance soil carbon uptake enough to make a difference? One way to answer this is to calculate the time for this enhancement to make a discernible difference, which I arbitrarily but reasonably define as 10% of the respective pools. The calculation shows that enhanced global soil carbon uptake by better grazing management can remove 10% of the atmospheric carbon pool (90 billion tons) in 150–900 years, and add 10% to the soil storage pool

(30–55 billion tons) in 50–550 years. The upper bounds are impractically long, but the lower bounds, 50–150 years, are of interest. Our verdict is thus mixed: possible, but, with modest odds of significant benefits. We need a finer grained analysis.

Common Versions of the Grass-Grazing Argument

To quantitatively examine the view that beef grazing benefits climate, it helps to distinguish two variants of the view, a narrower and a more sweeping one.

The narrower variant addresses only one phase of beef production, finishing, during which weaned (typically six- to ten-month-old) animals gain the final 50–65% of their slaughter weight in a relatively brief sprint.[381,571] Conventionally, finishing relies almost exclusively on concentrated feeds and fine hay and silage. These use only high quality cropland, not rangeland, as well as fossil fuel–based, CO_2-intensive machinery and agrochemicals (synthetic fertilizer, herbicides). Accepting as an axiom the notion that cattle grazing enhances carbon sequestration, and that this putative enhancement can fully offset production emissions and then some, the narrower variant sets out to reduce beef's CO_{2eq} emissions by replacing intensive, high-CO_2 finishing with presumed carbon-negative grass finishing.

The broader variant does not address a specific phase of beef production, but focuses on grassland management models—variably called "rapid rotational" or "adaptive multi-paddock" grazing[413,589,591]—in which cattle rotate rapidly among paddocks, briefly and intensely grazing each. Proponents of these models believe that managing grasslands according to the models' dictates can greatly enhance sequestration, enough on average to offset beef's large production emissions and render cattle grazing a general carbon sink[248] that measurably reduces anthropogenic climate change.

Both variants are observationally and logically strained. Their basic challenge is rangelands' low productivity, which results in rangeland-based finishers taking much longer to reach market weight. In one not particularly extreme example,[92] while grass fed finishers took 679 days to reach a final live weight of 486 kg, their conventional counterparts reached 571 kg at a final age of 440 days; 20% higher final weight, in 35% less time. In terms of the all important average weight gain rate, the grass fed finishers' was 610 g per day to the conventional finishers' 1,860; more than threefold faster weight gain by the conventional finishers.

This situation has two downsides. First, the market is disinterested in chewy beef of old steers. More importantly, long-living finishers require more feed, resulting in more than emission doubling, from about 180 kg CO_{2eq} per kg of produced beef protein in herds with high intensity finishing to about 380 kg CO_{2eq} per kg of produced beef protein in rangeland-based herds.[177] Because these higher feed needs raise production costs significantly, most grass finishing uses cropland, often aided by irrigation and agrochemicals. The narrower variant thus still uses resources[456,571] that could equally well produce almost any alternative food item. In terms of alternative resource uses, therefore, advocated grazing-based beef finishing is most logically compared to producing either energy- and protein-dense feed for intensive livestock operations or plant-based food for direct human consumption. Based on the body of solid research we have already partly encountered, the human food route would feed more of us, more efficiently, with better environmental and public health outcomes.

Because genuine rangelands can produce no human food, when rangeland based, the broader variant must be reasonably weighed against rewilding of the used rangeland.[79,583] This has an added wrinkle, however, because virtually all beef uses at least some cropland-based grain, hay, and silage. Every rangeland hectare freed from producing beef therefore has a corresponding freed cropland area that can be repurposed to producing food for direct human consumption. Since beef uses roughly 9 ha of rangeland for each cropland hectare used,[179,180] forgoing a beef mass that currently uses A ha entails simultaneous rewilding of $0.9A$ ha of rangeland and repurposing for human food production $0.1A$ ha of cropland.

Both variants make rudimentary biogeochemical sense, because grass photosynthesizes, photosynthesis takes up and fixes atmospheric CO_2 (turning it into radiatively inactive belowground liquid or solid forms), and most soil organic carbon indeed derives directly or indirectly from the aboveground biomass this photosynthesis produces.[14] But beyond this basic description, the viability of the above variants is determined by *rates* of these processes with or without cattle grazing, not their existence, which is not in question. Past this point, therefore, things get less obvious.

First is the link between aboveground living vegetation and organic litter, and soil organic carbon. The longer dead biomass from local vegetation spends as litter on the ground, the longer it has to react with atmospheric oxygen, decompose, and remineralize, thus returning its stored carbon to the atmosphere as radiatively active CO_2. Trampling and manure deposition by grazing cattle thus become important, because they short-circuit this process, bypassing the litter phase and delivering aboveground living biomass directly into the soil carbon pool.[41,557] Lowering losses of primary production to litter decomposition by the manure pathway is a clear benefit of cattle rumination, arguably cattle's most spectacular evolutionary trick (but rumination also elevates methane emissions). This simple carbon mass balance argument makes for a plausible conjecture—that cattle grazing may promote belowground carbon sequestration—that is supported by such mechanisms as enzymatic response of soil microbiota[500] or biochemical acceleration of plant regrowth.[676] To produce carbon-desirable beef, any "improved" grazing practice must robustly accelerate sequestration well beyond natural rates, enough to offset beef's high operational emissions and yield sequestration-corrected beef emissions that are comparable to or preferably lower than those of alternative foods.

Unfortunately, these expectations almost never translate into measurable reality.[377] The evidence, key highlights of which we have already encountered, can be summarized as follows (with more details in the Accompanying PDF). Soil sequestration estimates vary widely, especially with sites' long-term mean temperature and precipitation. Almost all high estimates (exceeding about 250 kg C per hectare per year) are observed in lush sites, and almost all low estimates (under 100 kg C per hectare per year) are obtained in semi-arid or drier ones. Since the two precipitation regimes roughly correspond to grazed croplands and rangelands, respectively, sequestration rates exceeding about 300 kg C (1,100 kg CO_2) per hectare per year are observed almost exclusively on lush croplands, not rangelands. When using rangelands, grazing by grass fed beef thus sequesters too little additional carbon to offset its high production emissions. And on cropland, where added sequestration rates are arguably sometime high enough, grass fed beef use land that can produce other food items than beef, thus still constituting a net carbon loss relative to the relevant viable alternatives, because when repurposed for those alternative foods, the land yields more, nutritionally superior food whose carbon emissions per kg protein are well below the sequestration corrected ones of grazing beef.[177]

Let's examine some key analyses that form the basis for this summary. An international review by Aryal[21] synthesized 151 studies published over 1993–2022 from a wide range of worldwide locations and conditions that address grazing effects on functioning and carbon storage of widely varied grasslands. The analysis focused on vegetation composition, primary productivity, soil fauna, soil physical properties, topography, and erosion rates. To resolve the roles of grazing, all results were reported as functions of grazing intensity, a measure of the density of animals on the landscape and thus the burden they exert on the ecosystem (e.g., kilograms of eaten grass per hectare per year, say). If you conjure up your favorite western, grazing intensity is the average number of wild horses or bison that occupy a hectare of rangeland on your screen. In Aryal's analysis, grazing intensity ranged over 0.04–0.45 livestock units per hectare. (A livestock unit is an imprecise measure of the average nutritional needs of one low-productivity mature beef cow and her unweaned calf.) The above intensity range roughly corresponds to an acre of rangeland sustaining a typical cow–calf duo for 6–65 days per year (requiring about 6–60 acres per duo for the whole year). The synthesis' bottom line: heavy, moderate, and light grazing intensity reduced soil organic carbon in 74%, 40%, and 14% of the papers, and increased it in 7%, 17%, and 36%. In other words, the worldwide reality the Aryal[21] synthesis revealed is mostly: the more grazing, the *less* organic carbon soils hold, exactly the opposite of the *enhanced* sequestration that grass fed beef advocates claim. Note that in the Aryal synthesis, under light grazing, more papers reported adding soil carbon than losing it relative to no grazing (36% vs 14%). While this finding suggests that light grazing does add carbon to the soil, this is contradicted by the broader global analysis of Ren et al. mentioned earlier,[486] which comprised seventy-three comparisons of light grazing to no grazing within latitude 30°–50°N, mean annual temperature above 1°C and elevations below 3 km. Of those seventy-three comparisons, in forty-two papers—58%—soil carbon storage was lower under light grazing than under no grazing. Both the mean and median of those seventy-three pertinent comparisons were negative, -1.3 and -0.4 ton per hectare. And there is more: while eighteen comparisons feature differences above $+1$ ton per hectare, fifty-five differences—threefold more—are below -1 ton per hectare.

Negating such disappointing results, some analyses suggest that the grass finishing variant may indeed enhance the environmental performance of beef. A particularly compelling example is the careful Pelletier et al.[456] modeling study mentioned earlier of Midwestern US beef production. It shows that whereas feedlot and grass-finished beef require emitting about 15 and 19 kg CO_{2eq} per live kg of beef produced, assuming generous carbon sequestration on the utilized cropland-based grasslands reduced these values to 13 and 11 kg. Including carbon sequestration by the grazed cropland thus turns grass finished beef from being 30% more to 15% *less* greenhouse gas intensive than conventional finishing.

But let's probe these results more carefully. To cast beef in the most positive light possible, let's focus on their lower bound of sequestration corrected net beef emissions, 11 kg CO_{2eq} per live beef kg produced, or—assuming 43% boneless fraction and 220–260 g protein per kg beef—100–120 kg CO_{2eq} per kg beef protein produced. For the following discussion, keep in mind that this is exceptionally low, about half of our 2014 data-based estimate[179] for US beef, and one fifth of the mean of an unusually comprehensive global analysis of beef emissions:[475] i.e., presenting an exceptionally positive case for beef.

Yet even this unusually low estimate of sequestration corrected grass finished beef emissions is much higher than the corresponding emissions of food alternatives, as we'll cover soon: roughly two, four, and ten times those of poultry or farmed fish; grains; and groundnuts, peas, or legumes.[475] These comparisons are pertinent and fair because the Pelletier herd grazed on cropland, not rangeland (which is essentially absent from the area they addressed, the upper Midwest). So, calling beef more desirable than these alternatives based on carbon balance is mostly false, even when the deck is stacked in beef's favor by using the low-end estimate of its carbon intensity. It is true that after correcting for sequestration, grass finished beef production requires slightly lower emissions, 100–120 kg CO_{2eq} per kg beef protein produced, than conventionally finished beef, 120–140 kg CO_{2eq}. Yet the average emissions per kg protein of nuts (pistachios, almonds, walnuts, hazelnuts), legumes (soy, lentils, kidney beans, and chickpeas) and cereals (spelt, barley, buckwheat, wheat, and sorghum) are 3–4%, 8–10%, and 18–21% of conventional beef's. Even chicken and eggs are 27–28% of beef's emissions, and pork's are 65%. Some may celebrate the 15% emission edge sequestration endows grass finished beef with relative to conventional beef. But rationally, this is rather trivial by comparison with the fact that emissions by nuts, legumes, and cereals are five to thirty-five *times* lower than conventional beef's.

The Cattle as Latter-Day Bison Objection

Addressing methane dominance over total CO_{2eq} emissions of beef, especially when grass based, some beef proponents invoke presumed parallels between current-day cattle and American bison that roamed North America centuries and millennia ago. How are modern cattle methane emissions different, they ask, from those of bison, with which Earth's climate was in perfect harmony for millennia? Why are the same burps an environmental time bomb when coming from cattle, but just a natural process when coming from bison?

Because what matters is not who does the burping, but the climate in which the burping is done. And Earth's climate in the last century, when human activity has been raising the concentrations of the dominant anthropogenic greenhouse gases (CO_2, methane, nitrous oxide), has been completely different from that between the end of the last glacial period some 12,000 year ago until roughly 1750. Atmospheric CO_2 concentrations were about 180–190 parts per million by volume (ppmv) at the end of the last glacial period, rose naturally to about 280–290 ppmv between then and the industrial revolution, and have been rapidly anthropogenically rising since, exceeding 430 ppmv in Alaska in April 2024.[610] Similarly, methane concentrations were 350–370, 580–670, and 1,925 ppbv (parts per billion by volume)[610] in the respective times, a rising trend also exhibited by nitrous oxide.[610] And along with this atmospheric greenhouse gas buildup, closely tracking the theoretically expected response based on the physics of the anthropogenically enhanced greenhouse effect, global mean surface temperatures rose between the last glacial maximum and about 6,000 years ago, declined gently until roughly 1800, and have been rapidly rising since, in January–September 2024 approaching 1.3 K (1.3°C or 2.3°F) over preindustrial temperatures.[276,446]

The message of these observations is clear: while during the Holocene reign of bison as the principal North American grass grazer the climate was minimally anthropogenically impacted, and was marching to essentially orbital forcing plus modest human perturbations (chapter 1), now the anthropogenic impacts completely dominate the climate response,[276] altering not only Earth's surface temperature but indeed the climate system's modes of internal variability.[128] Just as extreme heat or prolonged exertion affect differently a twenty-five-year-old endurance athlete and a seventy-five-year-old congestive heart failure patient, it should be entirely self evident that the same effect—ruminant emissions—would have two entirely distinct impacts at two distinct time periods—the Pleistocene through the mid Holocene vs the twentieth to twenty-first centuries—during which the climate system is under completely unprecedented level of human induced stress.

The above is further emphasized by the fact that the ruminant emissions in question are dominated by methane. Because its lifespan in the atmosphere is brief, on the order of a decade, methane atmospheric concentration reflects a balance between rates of emissions, oxidative conversion to CO_2, and concentration changes with time if the two are significantly out of balance. From the end of the last ice age until the industrial revolution, emission rates fluctuated minimally, in response to natural variability (of, e.g., the extent of tropical wetlands) and to changing human numbers, technology, and customs. Methane concentrations and warming were thus roughly steady, reflecting an approximate cancellation of the emissions by oxidative disappearance. Since now atmospheric methane concentrations rapidly climb, reflecting rising decadal mean emissions (from all sources, notably including fugitive emissions from gas exploration), methane warming is *not* fixed, but rather rising. This makes the effects of current methane emissions entirely distinct from those of the mid Holocene.

Our anthropogenically warmed present climate is thus largely not comparable to that of centuries ago. By highlighting the current dominance of anthropogenic climate change, the match between atmospheric greenhouse gas concentrations and global mean surface temperatures makes clear that the only way to address climate change is to eliminate and eventually reverse atmospheric buildup of *all* greenhouse gases, including methane.[462] The bison objection is therefore logically flawed, because it ignores the dramatic differences between the two relevant time periods: roughly stable or very slowly varying atmospheric composition and surface temperatures during the millennia preceding the decimation of Plains bison populations, but rapidly rising now. New realities, new view of greenhouse gas emissions; despite methane emissions from historical Plains bison posing no problem, methane emissions from modern cattle pose an imminent threat,[545] and must be immediately curbed.

Beyond the conceptual differences described above, additional pertinent numbers pour further cold water on bison fantasies. In 2021, US beef production emitted at least 140 million metric tons CO_{2eq} (obtained by combining USDA data[439] with data from the US-EPA Greenhouse Gas Inventory Data Explorer). A more expansive estimate,[498] of early 2020s life cycle annual emission related to beef production and consumption in the US is about 250 million metric tons CO_{2eq}. Are these emissions comparable with historic bison emissions? To derive an upper bound on bison emissions, let's multiply the highest estimated peak North American bison population, 60 million[124,307] by a high estimate of individual bison annual methane emissions of 60 kg CH_4 (1,680 kg CO_{2eq}), yielding emission by the bison herd of 100 million metric tons CO_{2eq} a year. Modern beef methane emissions are thus at least 40% larger than and quite possibly more than double historic bison emissions, and those additional emissions enter an atmosphere already bloated with methane, in marked contrast with the reign of the bison era. And methane is of course not alone in beef's current emissions, which also include CO_2 and N_2O, but was mostly alone in bison emissions (in reality we must assume there were also some manure-related nitrous oxide emissions). Like the grass feeding advocacy, the bison objection to the characterization of beef as resource intensive is at odds with reality, and should persuade nobody.

Promoting grass fed beef as carbon beneficial because grazing enhances sequestration raises another, subtler methane-related challenge that is poorly captured by comparing competing scenarios' total emissions, summed problematically into CO_{2eq} emissions (defined in chapter 1). The issue arises because methane emissions of a given cow or steer are higher on grass rations than on concentrated feed, because grass contains more fermentable structural carbohydrates. For example, a 400 kg steer may emit over 1.1–1.6 kg CH_4 per week on grass, but only about 0.6–0.8 kg a week on intensive rations. [171,235] On the other hand, lower CO_2 emissions are incurred in producing a beef calorie from grass than from concentrated feed. Overall, therefore, grass feeding *may* lower slightly total CO_{2eq} emissions, but is sure to raise the methane-to-CO_2 ratio in total emissions, while concentrated rations lower it. This raises the following questions.

Because of our sustained failure to address anthropogenic climate change, we only have several decades left to finally address the issue. [271] On these short timescales, methane, the quintessential double-edged sword, becomes central. On the one hand, once emitted into the atmosphere, methane rapidly converts (oxidizes) into CO_2; of a kilogram of methane emitted twenty years ago, less than 0.2 kg remains in today's atmosphere as methane. Methane's big plus is that its brief atmospheric lifespan makes the warming by a single pulse of methane mostly reversible; whereas CO_2 continues to warm Earth's surface many millennia after it was emitted, the methane pulse does so appreciably only for a few decades. The minus is that during this time, methane that does exist is very potent, warming Earth's surface roughly a hundred times faster than CO_2. [536]

Let's reexamine grass fed beef in light of what we've just learned, momentarily suspending judgment and supposing at face value that its net CO_{2eq} emissions are indeed smaller than industrial beef's. Then, favoring grass fed beef would reduce CO_{2eq} emissions somewhat, but raise the methane-to-CO_2 ratio (due to anaerobic grass fermentation). This means accepting a higher surface warming in the near term, before methane oxidizes into CO_2, in return for reversibility of the warming if beef consumption were to drop markedly at some point. In that case, atmospheric methane would decline rapidly to lower steady-state concentrations, eliminating most of the high initial warming. Crucially, this "grass fed beef and high warming now, something will surely work out later" option partially liberates us from the commitment to the initially lower yet "essentially forever" warming that CO_2-dominanted conventional beef entails.

Is this a wise choice? If you consider beef an optional food item, readily interchangeable with such alternatives as poultry or pork, not to mention lentils or chickpeas, then in almost all cases absolutely not (with at least one hypothetical exception[175] that is too esoteric to delve into here). If you deem beef a vital necessity instead, and you have rigorously identified a beef independent way to maintain Earth climate within a range compatible with human societies in coming decades, then possibly yes, *on average*, i.e., if surface temperature were rising linearly with greenhouse gas concentrations, steadily warming at a fixed rate as greenhouse gases build up in the atmosphere. While with the benefit of a few centuries of data or less Earth may appear to mostly behave this way, this may not always hold, and it may be a beguilingly reassuring feature we may not wish to rely on. One reason for this is that it's becoming more and more likely that Earth features, at least in principle, the deeply concerning possibility of practically almost irreversible "tipping points" that the climate system may near.[196] Such points, or regime changes, can be heuristically envisioned by imagining warming water. If you start from 15°C, say, and invest enough energy in the water to warm it to 20°C, and then invest the same amount of energy again to further warm the water, you will get water at approximately 25°C; things seem stately and well behaved. But now imagine you start instead at 96°C, again investing the same amount of energy in further warming the water. At some point (when the water is at 100°C, the boiling temperature of water), a transition point is reached; instead of liquid water, you now have water vapor. The regime has shifted, from one in which water is liquid to one in which it is gas. Let's use a specific example—the so-called Sea ice-Albedo feedback—to examine putative climatic tipping points.

A basic intuitive expectation of Earth's climate is warm tropics and cold poles. However, robust, diverse evidence illuminates times in our geological past during which crocodiles, palm trees, and other warmth-loving life forms lived on Arctic Ocean islands that dotted an ice-free open ocean.[123,579] The physics of this foreign, "crocodile friendly" Arctic may have arisen dynamically[189] (involving shifting of planetary scale wind patterns), but likely also involved the transfer of electromagnetic radiation through the atmosphere, and thermodynamics. Today, these processes are key to keeping the Arctic ocean cool: sporting an expansive sea ice cover, the ocean stays cool because most (at least 80%) solar energy hitting its icy surface is reflected skyward before its absorption by the surface, or warming, can occur. But notice the circularity of this process: sea ice is prevalent in the Arctic because even at local summer noon, the sun in high latitudes is never very far above the horizon, resulting in minimal solar warming by oblique incoming solar radiation. This results in a cold surface, which permits sea ice. But that coolness is furthered by the high reflectivity of sea ice. This is a classic example of a positive feedback loop, in which a process—solar surface reflectivity—alters the basic state of the system in which it operates—Arctic temperatures—so as to further its own perpetuation, keeping the Arctic cold, iced over, and thus highly reflective. A climate-unrelated example, starting a fire in a wood-burning stove, may clarify this further. You need a chimney updraft to kindle the fire, but an updraft requires the presence of intense heating, for which you need an established fire. But once some initial fire is going (by using paper, dried bark shavings, or whatever highly combustible, low heat capacity material you use to start your fires), an updraft quickly establishes, and more combustion proceeds apace. A process—here updraft—impacts the basic state of the system of which it is a part—the stove fire—so as to self-perpetuate (guarantee the conditions that give rise to the updraft, a hot stove interior). As we'll explore later, but in broader mechanistic terms, a similar positive feedback loop holds, in reverse, in deserts.

Let's quantify the effect of the sea ice feedback. Robust sea ice reflects about 80% of incoming solar radiation. Summer daytime incident solar radiation typical of the Arctic surface is about 300 watts per square meter, which means that the sun delivers 300 joules of energy each second to every square meter of Arctic surface. But because only 20% is absorbed, the surface is only warmed by 60 watts per square meter. In contrast, the open ocean would reflect much less, say 20%—absorbing 240 watts over the same area. The fact that the sun warms the Arctic surface four times faster in open ocean than under sea ice cover gives rise to a potential tipping point that nicely exemplifies these points writ large.

The trouble begins when something—say an enhanced greenhouse effect through our emissions, or enhanced volcanic CO_2 degassing—works to slow Earth-to-space energy flux, raising surface temperatures. If no compensatory mechanisms counter the warming, sea ice would start disappearing in response. The warming rate will then not only be related to that external warming mechanism, but that warming rate plus that produced by the added solar energy now hitting the unshielded sea surface (more details in the Accompanying PDF). And rather than being fixed, this added energy flux itself keeps on rising with time, because every bit of newly ice-free ocean would enhance surface solar absorption, accelerating further sea ice retreat. This process would thus drive the Arctic inexorably and ever faster toward a warm, ice-free state (as quantified in the Accompanying PDF). And accompanying such an Arctic is a northern hemisphere starkly warmer and weirder than today, with impacts as far south as Florida, the northern Gulf of Mexico Coast, or the Mediterranean basin.[115,119,448,549] The transition from the familiar state, with the Arctic extensively covered by multiyear sea ice, to this wildly different climate may well qualify as a regime shift,[369] a transition of Earth workings from one state to a qualitatively different one.

For humans and our societies, the key point in the discussion on tipping points is that the historic state being replaced here is the climate with which we evolved, first biologically and—more recently and impactfully—culturally. Some impacts of this regime shift seem decidedly pedestrian. For example, in New York State, while three-inch gutters were historically normal and adequate, new houses must now use, by code, five-inch gutters, in recognition of the proliferation of once exceptional precipitation events into the new norm.[550] Other impacts are existential, including the potential tragic collapse of once thriving human civilizations so utterly defeated by their changing environments that the only options left are en masse emigration or death. Such catastrophes are rarely propelled by only deteriorating climate, or even environmental conditions more broadly. Instead, they often culminate a blend of rapid severe climate change[108,306] with other societal forces, such as fraying social cohesion, poverty, or sectarian strife, which jointly deprive citizens[361] of the ability to organize effective centrally planned protections from such climate change impacts as rising water scarcity or more frequent, extreme heat, especially at night.[320] Such tragedies—unmistakably told by the ghost dwellings of Mesa Verde[527] and the long abandoned early

Holocene human dwellings throughout the Arabian peninsula[342]—are sadly not uncommon.[150] In fact, as Vivian Yee reported in the January 19, 2023 *New York Times*,[664] a miniature rendition of this process is currently unfolding in southern Tunisia. In this desert locale, persistent droughts and the allure of distant and better economic prospects than the traditional sheep herding and olive growing now combine to end a unique local culture of cliff dwellers some 1,000 years old. Such is the potential power of climatic regime shifts.

If tipping points prove to exist and to have widespread impacts, as appears more and more likely,[22,77,243,256] our choices on beef—how much of it we eat and whether we grow it conventionally or on grass—could partly determine our climate future. For the coming few decades, during which this future may be determined, the conventional (CO_2 dominated) beef route offers less warming than the grass (CH_4 dominated) route. This *may* permit life to remain within the current climatic regime (envelope of expected fluctuations) in which we and our societies evolved. But the price of that smaller warming is that we commit to it (consider it an inevitable and acceptable price of beef) essentially forever, because its CO_2 dominance spells warming that endures for millennia.[16–18]

Conversely, because methane oxidizes to CO_2 on a roughly decadal timescale, the grass feeding option offers—uniquely and comfortingly— reversibility of the ensuing warming. This does not help if we maintain our current beef consumption because, in that case, atmospheric methane will rise until attaining a higher steady state in which additions by new emissions just match removal by oxidation, and in which the surface bakes, and fast. But it does matter if we drastically reduce beef consumption or learn to drastically cut its methane emissions. While neither seems likely at the moment, with either change, the last methane molecules emitted before the hoped for change would linger thirty to fifty years, beyond which no further methane-induced anthropogenic warming will occur. This is the reversibility comfort the grass model offers, but it too comes at a cost.

Yes, the grass option warming is reversible, but the rate of Earth warming it induces is also higher. And if our overall warming trajectory happens to approach a tipping point, then the extra warming due to favoring the grass option would tangibly elevate the odds of transitioning into a climatic regime that is a marked enough departure from our historic path to be incompatible with key aspects of modern life.[615] And once we transitioned, the road back into the familiar climatic regime within which humans developed requires emission reduction efforts far beyond the ones required now.[315,316]

So, if your primary motivation is to prevent climatic regime changes, you would logically favor conventional beef, lowering the odds of such changes. But in doing so, you would also sign on to an anthropogenically warmed Earth for a long, long time indeed. If, instead, you are less concerned about climate regime changes, but mostly about Earth's long term habitability, you'd favor grass fed beef. Because this option entails elevated methane emissions, it would inevitably mean very warm few decades. That is the cost. The benefit, assuming the warming is not so large as to propel a climate regime change, is that once we find a technological solution to beef's methane problem, or find it in us to lower beef consumption, within several decades we will return to minimal and manageable beef related warming. But what if this untested assumption proves untrue, and the added methane emissions do compel Earth into a new climatic regime? If this gives you pause, I can easily understand why. Favoring either beef option over forgoing beef altogether is therefore rolling the climate dice. It makes no rational sense without a robust demonstration that your choice environmentally outperforms both the other option, and—most pertinently—reasonable alternatives to beef. As we will see shortly, this is easier said than done.

Recall that the impetus for primarily grazing beef cattle is the hope— despite the observations summarized above, in which grazing mostly undermines sequestration—that grazing would enhance soil carbon sequestration, and that that enhancement would prove large enough to offset the emissions associated with beef production. Disregarding beef's other and considerable environmental liabilities (chapter 2), by how much must grazing enhance sequestration rates to justify this expectation? In the Accompanying PDF, I derive the threshold conditions for carbon neutrality, when cattle grazing *sustainably* enhances sequestration rates beyond ungrazed rates to fully offset all operational emissions. Using US and global data, I show that such neutrality is realized when sustained sequestration enhancement by cattle grazing is 250–6,000 kg C per hectare

OF BEEF AND PUBLIC CONFUSION, TAKE II: THE ENVIRONMENT

annually. Above, "sustained" is a critical qualifier, because much of global grazed lands are currently extremely carbon-depleted by decades of unsustainable exploitation by overgrazing, and thus often exhibit very large sequestration rates in the first few years after cessation of active abuse. Because such added rates are entirely transient and ephemeral, they wildly overstate the land's actual long-term sequestration potential.

Can grazing under better managed rangelands sustainably enhance annual sequestration by 250–6,000 kg C per hectare? The thorough—essential yet necessarily pedantic—survey of the relevant scientific literature on carbon sequestration modifications by grazing in the Accompanying PDF reveals that, as we have seen, these modifications can be positive *or* negative, and vary widely in magnitude. To give you a sense of how wild this variability can be, here are some estimates discussed in details in the Accompanying PDF, all in kilograms carbon per hectare per year, with positive denoting downward atmosphere-to-soil flux: +250; +130; +20–40; +300 and others with roughly zero mean; −155; a few sites with +250–300 and many with zero; a statistically unsound mix of +300, −300, and 0. Some authors favor reporting results in relative terms, organic C under grazing as a percent of ungrazed land, finding: 30–40% lower; 16–18% lower; 45% lower; 0%; 0%; and positive and negative effects on mineral associated soil organic matter content with a roughly zero mean. Two analyses of ungrazed plots show annual sequestration rates of 1,000 and 4,200 kg C per hectare, [608,661] demonstrating that grasslands can vigorously sequester carbon also when not grazed, countering the view of grazers as obligatory and indispensable to grassland functioning. If this tangle of numbers—positive and negative, with averages that mostly congregate around zero—doesn't quite lull you into believing that cattle grazing decisively enhances soil carbon storage, well, perhaps that's because the numbers just don't show that it does.... If we refer back to the very comprehensive meta analysis by Ren et al. we encountered earlier, [486] thirty of the studies the authors meta-analyzed also include the duration of the comparison, how long the light and no grazing conditions in the two members of the compared pairs have been going on. Since the compared plots in a given pair are mutually adjacent, we can reasonably assume that before the comparison started, the two plots' soil organic carbon stores were essentially the same. This permits deriving estimates of exactly what we need here: the annual soil carbon uptake enhancement by light grazing relative to no grazing,

$$\text{added sequestration} = \frac{\begin{array}{c}\text{soil organic carbon under light grazing}\\ \text{– soil organic carbon with no grazing}\end{array}}{\text{years of distinct grazing management}}.$$

Using the Ren et al.[486] data to take advantage of this possibility, we get overall mean and median of the thirty sites with the necessary information of about 70 and 40 kg C per hectare per year; positive, but very small. Moreover, the average in sites in which light grazing enhances sequestration relative to no grazing is 470 kg C per ha per year, while the average in sites in which light grazing *reduces* sequestration relative to no grazing is −730 kg C per ha per year.

When shoehorned into a statistically strained geographically representative "best estimate" of grazing sequestration enhancement, these decidedly mixed observations indicate mostly modest increases in soil carbon storage. Enhancements of 0–200 kg C per hectare per year are possible and not unlikely, but larger enhancements—while possible—are decidedly the exception.

Notice that the 0–200 kg likely range is below even the lower bound— 250 kg carbon per hectare per year—of the breakeven sequestration that is required for carbon neutrality. As the Abstract of one of the above compilations[561] succinctly put it, the results are saddled with "high uncertainty," not due to errors or sloppy scholarship, but due to the great variability and multifaceted biogeochemical and physical dependencies that govern atmosphere–soil exchanges. This is echoed by the caption of Supplementary Online Figure 4 of[228]: "This figure highlights that the range in estimates is large, which reflects the uncertainties inherent in developing estimates of this type, but also differences in the management practices considered and associated definitions, the geographical and agro-ecological focus of the studies and methods of data acquisition or generation including the time frame over which the sequestration rate is averaged."

Although this chapter has introduced several compelling reasons for doubting beef grazing's environmental benefits, this discussion is only as robust as our currently decidedly imperfect quantitative insights into the biogeochemical life of rangelands. Given how varied and vast rangelands are, the multitude of ways we meddle with their affairs, the complexity of their biogeochemistry, and the fact that much of this complexity—while intimately coupled to surface processes—unfolds below the surface, inaccessible to unaided eyes or most remote sensing platforms, this partial ignorance makes perfect sense. To the extent permitted by this imperfect knowledge, we can say that grazed rangelands can be net carbon sinks or sources; that when they are not actively abused or overused they are more likely to be sinks than sources with or without grazers; and that the net carbon uptakes of true rangelands, as opposed to grazed croplands, [413] are mostly in the 0–200 kg C per hectare per year range. For example, a compilation of sixty-seven published paired comparisons of soil organic carbon stocks in grazed to ungrazed sites [466] revealed widely divergent grazing effects on soil organic carbon, with increases, decreases, or no change. Focusing then on the relevant precipitation range, 400–850 mm a year (below which productivity is too low for expecting high sequestration rates, [321] and above which the area is more likely to be cropland rather than rangeland) revealed that in nearly all comparisons within this range, root mass, a key component of soil organic carbon, either decreased or did not change. The portion of the mostly small carbon uptake by grazelands uniquely attributable to cattle grazing and its dependence on a particular grass management practice are even more uncertain. [122]

Meaningful mitigation of beef cattle's greenhouse gas emissions by grazing is most likely where operational emissions are relatively low and sequestration potential is high. This singles out lush grasslands, such as the Midwestern grass-finishing operation Pelletier analyzed, [456] as the prime candidate for such offsetting. But in most such landscapes, this small putative additional carbon sequestration is a secondary red herring at most. In these environments, water availability, topography, and other determining physical factors permit production of greater amounts of alternative, more nutritious foods. [183]

Is Grass Finishing Environmentally Wise?

As we encountered earlier, producing beef requires roughly twice the greenhouse gas emissions per kilogram that protein poultry or farmed fish do, four times what grains do, and ten times the emissions peanuts, peas, or other legumes require.[475] Possible yet uncertain carbon sequestration enhancement by cattle grazing[430,431,456,563] changes matters minimally or not at all, because the magnitude of this putative enhanced carbon sequestration by grazing is most often, in most locations, overstated.

To further illustrate this wide disparity in resource needs, let's compare the numbers for the production of 1 kg of protein from grass fed beef and alternatives. For the needs of grass fed beef, let's use again the estimates of the Pelletier et al. analysis[456] (their Figure 1, assuming again 43% boneless fraction and 220–260 g protein per kg beef): 170–200 kg CO_{2eq} and 1,070–1,270 square meters of land per kg beef protein produced. We need not distinguish cropland from rangeland in this comparison because all the land Pelletier et al.[456] considered is cropland used for grazing, not rangeland. For the corresponding needs of alternative protein sources, let's use the classic Poore and Nemececk 2018 compilation.[475] Averaging over all non-beef food classes in that compilation, producing a kilogram of protein requires emitting 35–45 kg CO_{2eq} and 65–80 square meters of cropland. This disparity—170–200 kg CO_{2eq} and 1,070–1,270 square meters of cropland per kg beef protein vs 35–45 kg CO_{2eq} and 65–80 square meters per kg alternative protein—distills the stark choice in front of us. Option one is to cling to a small amount of grass finished beef, the quintessential "putting all one's eggs in one basket" approach. Option two is to partition the land and related resources between some production of a mixture of alternative crops, from which more bountiful, diverse, and nutritionally superior diets can be made, and some land sparing and rewilding.

To clarify this important conservation opportunity, imagine a cropland-based grass fed beef operation of the type the Pelletier analysis addressed, occupying 100 hectares (roughly 250 acres). Given the resource needs of grass fed beef enumerated earlier, this operation produces about 790–940 kg beef protein and emits 135–190 metric tons of CO_{2eq} per year. Let's imagine the owners of such Midwestern beef operations enacting a radical change: sparing and rewilding 85% of their cropland (85 hectares each). If replicated widely and done thoughtfully, this would reduce the carbon footprint involved, and afford more area for biodiversity (quantified shortly). Beyond the sheer area this option bequeaths to wildlife, with smart spared land geographical architecture, this would also provide contiguity of wildlands that is so essential to species survival on these newly added habitats.[216] Such a collective action–fueled change would constitute major and badly needed tailwinds in the sails of the currently losing battle of biodiversity conservation.[168,421]

Let's now quantify food production and resource use of option two, and compare it to option one. For that, let's imagine the owners reallocating the remaining 15 hectares of "unwilded" cropland per operation to producing some mix of beef and assorted alternative crops whose resource needs Poore and Nemececk address.[475] Our bottom line per 100 cropland hectare operation is:

| | | | | per year | |
| Choice | cropland hectares | | | produced protein, kg | emissions, metric ton CO_{2eq} |
	rewilded	fallow	cultivated		
grass fed beef	0	0	100	790–940	135–190
non-beef alternatives	85	0	15	1,850–2,400	65–110
non-beef alternatives	85	5	10	1,250–1,600	45–70

Based on these simple environmental metrics alone, and ignoring the substantial public health benefits of the alternatives over beef, what rational land manager would favor the beef option over either alternative?

The more learned among beef proponents may object to this comparison by invoking the tragic fate of cultivated grasslands in the US southern plains during the manic "plow-up" that ushered in the Dust Bowl catastrophe.[166,654] Since the above "alternative" scenarios entail, in some geographies, the same abuse of lands that were never meant to be used for crops, is Option 2 not actively courting the next Dust Bowl? These beef proponents are right that, ideally, such grasslands would indeed never be tilled.[408] But feeding ourselves we must, leading to some relaxation of this sweeping prohibition, subject to an unyielding commitment to soil conservation. This balance is permitted in the "alternative" scenarios because they use only 15% of the available cropland yet produce 2–3 times as much protein. This affords more than ample flexibility for periodic fallowing,[2] increased reliance on perennials,[528,674] and the use of soil sparing techniques that have been developed since the 1930s Dust Bowl.[295,558] The bottom row of the table, for example, depicts the situation if we allow a full third of the cultivated land to be fallow at any given year (i.e., fallowing or cover cropping any field every third year). Even with this stringent soil conserving criterion, the "alternative system still yields 135–200% of the protein the beef option does while saving 50–75% of the greenhouse gas emissions. This is why the "alternative" option not only avoids the Dust Bowl mistakes but is indeed their antithesis.

But the above only considers cropland. Can the use of true rangelands instead overcome this emphatic inferiority of the allocation to beef? While these lands feature their own unique and somewhat uncertain calculus, using them for cattle grazing is also by no means a reproducible, robust atmospheric carbon sink,[95,263,385,455] and using them for finishing is in most cases practically impossible because (*a*) finishers on such meager rations reach market weight far too old for marketability, as old as three or more years old at slaughter time; and, more importantly, (*b*) such finishers are inefficient. We compared[177] the feed intake per kg beef protein produced by the same herd, one with the finishers relying on meager rangeland based rations (with metabolizable energy density of 1,900 kcal per dry kg) and the other enjoying rather rich rations (with 2,600 kcal metabolizable energy per dry kg). Because this is expressed as kg dry feed per kg beef protein produced, it takes note of the very different weight gain, lifespans, and feeding rates. Taking all of those into account, the rangeland and the rich ration herds required 320 and 180 kg dry feed per kilogram of beef protein. Rangeland-based herds thus require almost 80% more feed per unit output. This is what I meant by "inefficient" above.

The literature review of the Accompanying PDF makes clear that for our central question here—can better managed rangelands sustainably enhance sequestration by 250–6,000 kg carbon per hectare annually—the upper end, or even half of it, are out of the question; I know of no measurement of 3,000–6,000 kg carbon per hectare per year in actual rangeland beyond possibly the first few seasons of prudent land stewardship following long agricultural abuse. Conversely, the lower bound, 250 kg carbon per hectare per year, is uncertain but possible (as the Accompanying PDF and other studies[28,382,510] reveal). Yet it is clear that when it comes to carbon storage in soils of real rangelands, subject to the unforgiving economic constraints of capital centered developed economies, the ungrazed option is often the one to beat because in such conditions—which promote overuse, overgrazing, and agricultural abuse—grazing *reduces* sequestration.

There is one final objection beef advocates make to calculations like the above, which depict beef of any kind as resource-inefficient relative to non-beef alternatives. It recognizes and largely accepts the weight of the above evidence, but argues that beef production must continue, because grazers and grasslands are evolutionarily inseparable, grazing is essential to the ecological integrity of grasslands from lush prairie to arid steppes, and cattle are the only remaining ubiquitous grazer. I am skeptical of this. First, populations of some natural grazers are robust, even over-proliferated. Even if cattle were the only remaining ubiquitous grazer, shouldn't our land use deliberately help threatened natural grazers instead of displacing them? In addition, the notion that evolution has bounded grasslands and grazers in an inextricable obligatory relation is not without challenges, as described in the Accompanying PDF. Comparisons of grazed and ungrazed neighboring sites, the fossil record, and the analyses of grazing cessation all present a consistent picture: while grazed or ungrazed grasslands develop along distinct botanical trajectories,[650] both can sustainably thrive, and both can store organic carbon in their soils if not overstocked. Moreover, under the rules of the modern ranching game in wealthy nations, ungrazed grasslands often store as much carbon as most grazed rangelands do, and often much more.

Jointly, the above evidence indicates that cattle grazing mostly reduces somewhat carbon uptake by rangeland ecosystems, sometime enhances it modestly, and in few exceptional locations and times, enhances it markedly. Overall, the association of cattle grazing and rangeland carbon sequestration is poorly structured and underwhelming. Conversely, beef production's high resource intensity is firmly established, highly certain, and thoroughly vetted. Much of the evidence on various putative societal benefits of cattle grazing thus strains the limits of quantitative observational inference, and is far too ordinary to justify the extraordinary claim that beef is a "climate solution."[413]

5

An Alternative Look at the Public Confusion about the Environmental Impacts of Beef

The average American eats 5–7 kg of beef protein a year, amounting to an annual national consumption of 1.6–2.2 billion kg beef protein. Producing this beef requires[179,180] 6.2–8.8 million metric tons of reactive nitrogen, 53–74 million hectares of high-quality cropland (which yields fine cattle feed, but which can also yield any human food type), and 320–450 million metric tons of CO_{2eq} annual emissions. These needs amount to 52–74% of the total US use of reactive nitrogen fertilizer, 33–46% of the available cropland, and 5–7% of total emissions (figure 3 of[179]). One food, that contributes under 10% of US calories, thus dominates over US water pollution, fine cropland use, and agricultural production emissions (about a tenth of the nation's total emissions). If we set out to clean up our polluted streams and lakes, or reverse wildlife declines, not to mention reducing emissions, maintaining beef consumption at its current levels guarantees failure no less assuredly than leaving our transportation or industrial sectors unchanged does to combating anthropogenic climate change.

While these are basic, clear facts we have already quantified in myriad ways, like most facts in the food–nutrition–environment space, they have been challenged. While I find none of the objections to this almost inescapable observation persuasive, this chapter will give you the tools you need for being the judge. Rather than simply surveying yet again the objections, let's consider several agricultural vignettes with grass finished beef at their core, expanding the scope of conditions under which we quantify the cost–benefit balance of such beef and compare it to those of other beef types or alternatives. Recall that by focusing on grass-finished beef, the beef type most often promoted as not merely outperforming conventional beef but indeed environmentally advantageous, we give beef its best shot at environmentally shining. These comparisons, I believe, will expose the key weaknesses in those objections.

Let's focus on relatively lush grasslands, where most high soil carbon sequestration rates are observed (Accompanying PDF), because this stands to cast the most positive light on cattle grazing. These grasslands occupy topographically mild settings—the Midwest through the eastern Great Plains of North America, or the southern UK—with long and warm enough summers to permit high productivity, economically viable commercial grazing. Given their accommodating attributes, such grasslands are definitely not rangelands (Accompanying PDF), but are nonetheless important because they maximize sequestration.

With neither severe droughts, nor fires or punishing cold spells frequent enough to suppress woody vegetation, if left alone, such grasslands would gradually reforest. This raises the following question: Are we better off using relatively lush, flat, agriculturally high valued lands for producing grass finished beef, or is permitting some rewilding and reforestation where these lands naturally occur the better option?

Using Lush Grasslands for Beef Production vs Rewilding or Producing Alternative Food

Benefit 1: Beef Production

Lush cropland–based beef operations occupy about 90 m^2-y (90 square meters, occupied for a year) per kg live beef output,[456] which corresponds to yields of 11–13 kg beef protein per hectare per year. Sitting near the center of the range the classic Poore and Nemecek compilation spans,[475] 3–24 kg beef protein per hectare per year, 11–13 kg beef protein per hectare per year makes sense for the lush grassland system we address.

Benefit 2: Carbon Sequestration by Land Type

We need sequestration rates for lush cropland–based grasslands, where finishing takes place, and for the true offsite rangelands used for the birth-to-weaning, cow-calf phase.[456,593]

Modestly managed and recently converted, intensively managed lush grassland pastures sequester about 120 and 400 kg C per ha per year, respectively.[291,456] Assuming the cropland portion of the operation is evenly split among these management types, its mean sequestration rate is 260 kg C per hectare-year. Since this rate falls nicely within the range we derived in the Accompanying PDF, generously 150–300 kg C or 550–1,100 kg CO_{2eq} per ha per year, this is the range of cropland sequestration rates we will use in the following discussion.

Based on the extensive literature review in the Accompanying PDF and on its summary in chapter 4, considerably lower sequestration rates are clearly appropriate for true rangelands that are used in the birth-to-weaning cow-calf phase. Consistent with that literature review, and following the particularly relevant analysis by Li et al.,[343] let's generously assume true rangeland sequestration rates of 120–140 kg C or 440–515 kg CO_{2eq} per hectare per year.

Cost 1: Direct Operational Emissions Incurred During Production

An estimate of direct emissions incurred during production is

$$\frac{19.2 \text{ kg CO}_{2eq} \text{ (kg beef)}^{-1}}{0.43 \times 0.22\text{-to-}0.26 \text{ (kg protein) (kg beef)}^{-1}} \approx 170\text{--}205 \frac{\text{kg CO}_{2eq}}{\text{kg beef protein,}}$$

whose left hand numerator (top value) is emission intensity of grass finished beef,[456] and the denominator is beef's protein content per kg live weight (with 43% of live weight remaining after bone removal, and with 220–260 g protein per kg beef).

Cost 2: Land Needs

As mentioned above, the agricultural land needs of a lush grassland based US beef operation are about[456] 90 m^2-year per kg live beef produced, or 800–950 m^2-year per kg beef protein. This range is consistent with Poore and Nemecek's median (distribution midpoint) estimate,[475] 855 m^2-year per kg beef protein, and with the range spanned by six varied grass based beef operations in lush New South Wales, Australia,[490] 570–1,280 m^2-year per kg beef protein. While the US beef system as a whole has much higher total land needs, 3,300 m^2-year per kg beef protein,[179,180] this makes sense due to this system's heavy reliance on very unproductive western arid rangelands. We thus take the total land needs of lush cropland–based beef to be, expansively, 700–1,200 m^2-year per kg beef protein.

Because rangelands and croplands have distinct alternative uses, and they sequester carbon at different rates, we must distinguish the rangeland and cropland needs of our examined grass-finishing system. For that, we need this system's rangeland/cropland partitioning, which varies by geography and agricultural practices (e.g., soil amendments can raise yields and thus lower land needs). In Australia,[490] unimproved pasture accounts for 74–77% of total land needs. In grass finishing operations in the US Midwest,[456] the cow-calf phase, where cows raise their unweaned young mostly on actual rangelands, claims 65% of the total land needs. While for the whole US beef system rangelands account for 90% of total land needs,[179,180] we will discount this here because that system relies heavily on very unproductive western arid rangelands. Based on these observations, let's assume rangelands account for about 65–75% of total land needs of grass finishing operations.

With rangeland accounting for 65–75% of the total land needs of beef finished on lush grasslands, 700–1,200 m^2-year per kg beef protein (derived above), a kilogram of beef protein requires 450–900 m^2 of rangeland until weaning, and 250–300 m^2-year of cropland for grass finishing, respectively.

Benefit 1: Carbon Sequestration

The sequestration per kg beef protein is the product of two known quantities derived above: land needs per kg beef protein (noting that 1 ha = 10^4 m^2), and expected annual sequestration per hectare, for both distinguishing rangeland from cropland. This yields the following sequestration rates.

Rangeland: 450–900 rangeland m^2-year per kg beef protein × 440–515 kg CO_{2eq} per rangeland hectare per year = 20–45 kg CO_{2eq} per kg beef protein.

Cropland: 250–300 cropland m^2-year per kg beef protein × 550–1100 kg CO_{2eq} per cropland hectare per year = 14–33 kg CO_{2eq} per kg beef protein.

Total sequestration is therefore 35–80 kg CO_{2eq} per kg beef protein, to which croplands contribute about 41–42%.

The Resource Needs of a Typical Grass-Finished Beef Eater

For subsequent calculations, we need the resource needs of "an average American," who currently consumes 6–6.5 kg beef protein a year[299]. If finished on grazed cropland, producing this amount requires 0.27–0.59 and 0.15–0.20 hectares of rangeland and grazed cropland per person, respectively, and operational emissions of 1,000–1,350 kg CO_{2eq} per person per year. When crediting the beef operation for carbon sequestration on the cropland it occupies (i.e., when subtracting the sequestered CO_{2eq} from the emitted CO_{2eq}), this drops to 500–1,125 kg CO_{2eq} per person per year.

Exploring Alternative Options

The first option, the one that must be explored first because it is poised to yield the most benefits, is perhaps radical to some: forgoing beef altogether. This may seem odd, since we must eat, so simply forgoing beef without a suitable replacement appears to make little sense. Perhaps, and we will explore the environmental consequences of replacing beef protein with equal protein amounts from alternatives later. But, let's derive an upper bound on potential benefits by exploring this arguably radical option, motivated in part by the fact that the "average American" currently consumes[299] about 2,500–2,550 kcal a day, some 15–20% beyond the 2,100–2,200 kcal a day necessary for health and weight maintenance. Relatedly, while current actual per capita US protein consumption is 115–120 g protein a day, the average American adult weighs about 84 kg (about 185 lb),[209] and a person weighing that much only requires 65–85 g protein a day.[215,303]

Eliminating the 180–200 kcals or 16–18 g protein a day beef currently contributes to our daily food intake is therefore eminently possible, and—given the staggering 29.6 kg m^{-2} [209] Body Mass Index of that average US adult—a distinct improvement over the current state. If you agree that humans are better off reducing excess food intake, wouldn't it make perfect sense, as environmentally minded citizens, to eliminate the least sustainable element of our diet first? Let's evaluate the environmental consequences of this choice.

Alternative Option 1: Reforest the Lush Cropland, Rewild the Rangeland

Recall that the type of beef operations this discussion focuses on occupies both rangeland (for the pre weaning cow-calf enterprise), and fine cropland (for grass finishing) that would most likely reforest naturally if left alone. Our first order of business is therefore quantifying the consequences of this reforestation.

Allowing reforestation of the 250–300 square meters of cropland needed for annually producing a kilogram of beef protein would naturally take up atmospheric carbon. To estimate the magnitude of this benefit, we note that in temperate latitudes, forests hold about 60–90 metric tons more carbon per hectare than average grasslands[281,349,350] or more[162] (see figure 2 of [410] or https://toolkit.climate.gov/image/486). Assuming reforestation adds 60–90 tons of carbon storage per hectare over 75–100 years (until the new forest fully matures to a steady biomass) translates into uptake of 600–1,200 kg carbon per hectare per year. Note that this is likely an underestimate. For example, at Harvard forest, in Massachusetts, we and others found[176,609] roughly two- to fourfold higher mean carbon uptake rates for a century-old forest, indicating that the above is likely an underestimate. But let's conservatively go with 600–1,200 kg C or 2,200–4,400 kg CO_{2eq} per reforested cropland hectare per year.

Because this cropland sequesters 550–1,100 kg CO_{2eq} per hectare annually currently (when used for beef), the *added* sequestration the choice to replace beef with reforesting the cropland portion of the current beef operation delivers is 2,200–4,400 *minus* 550–1,100, or 1,100–3,850 kg CO_{2eq} per hectare per year. Over the cropland needed per kilogram of beef protein, 250–300 square meter, this amounts to added sequestration

AN ALTERNATIVE LOOK AT THE PUBLIC CONFUSION ABOUT THE ENVIRONMENTAL 79

of, conservatively, 30–110 kg CO_{2eq} per kg beef protein. Subtracting this benefit from the cost—annual operational emissions of 170–205 minus added sequestration of 30–110 kg CO_{2eq} per person—yields net (cropland sequestration corrected) carbon *costs* of 60–170 kg CO_{2eq} per kilogram of beef protein. Take note; net carbon costs, not benefits.

Because beef production and thus the cow-calf operation are discontinued, grazing the operation's 450–900 m^2-year rangeland per kg beef protein will cease. How this cessation would alter the carbon balance of these rewilded rangelands is not well known. The reality we summarized earlier is that for almost all actual rangelands today, grazing cessation enhances sequestration,[28] thus *adding* to the costs of favoring beef grass finishing over rewilding the rangelands and reforesting the croplands. But imagine, improbably, a rangeland in which cattle grazing adds to the background sequestration extra 150 kg C or 550 kg CO_{2eq} per hectare per year. Over the 450–900 rangeland m^2 the operation occupies per kilogram of beef protein, this sequestration reduces emissions corrected for cropland sequestration, from 60–170 to 10–150 kg CO_{2eq} per kg beef protein. Still, net carbon costs, not benefits.

Worse yet, in light of the Accompanying PDF, such sequestration rates are not often achieved in true rangelands, but are instead possible at some times, in some rangelands. Most importantly for the current discussion, even when they are realized, the contribution of cattle grazing to these sequestration rates is most uncertain, and in most real rangelands today negative. On average, therefore, reversing the 60–170 kg CO_{2eq} per kg beef protein net carbon costs of the operation via enhancement by cattle grazing of carbon uptake in the rangeland (cow-calf) portion of lush cropland–based grass finishing operations is, while not impossible, most unlikely. In the Midwestern herd Pelletier et al.[456] analyzed, for example, this potential offsetting did not materialize, and the grass-finished beef variety remained a net carbon source, albeit a smaller one than its conventionally finished counterpart.

Nor did it happen in a recently modeled semi-lush ranch in Texas.[314] With mean annual precipitation of almost 600 mm, the addressed location is definitely not arid or even semi-arid. In fact, most rainfed agriculture in Israel enjoys this much or less rain. While used for grass-based beef today, the ranch thus occupies cropland that can be readily split among rewilding of most of the area while producing other animal- or plant-based food types on the remainder, as we will quantify shortly. Because such semi-lush lands tend to sequester more carbon than true rangelands, the following analysis is favorable to beef.

Unfortunately, what we need—an estimate of the modification of carbon sequestration rates by cattle grazing—is not evaluated by the model, and is in general poorly quantified, as mentioned earlier. But the reported results do offer an indirect proxy for the magnitude of this modification, the responsiveness of carbon sequestration in the ranch to management decisions. It is evaluated by comparing soil organic carbon under "the optimal" grazing method and other practices. (We will revisit the "optimal grazing practice" discussion later.) Under all grazing practices, soil organic carbon at the ranch increased for about two decades, and declined subsequently, at different rates (right panel of figure 4 in 314). At the fortieth and final year of the analysis, the soil contained 600 and 2,600 kg carbon per hectare more under the "optimal" method (their table 2) than under lightly and heavily stocked continuous grazing. Over forty years, such differences mean that the "optimal" practice resulted in taking up 15 and 65 kg carbon or 55–240 kg CO_{2eq} per hectare per year more than the two inferior methods. Since the latter range is far smaller than the 550 kg CO_{2eq} per spared rangeland hectare per year we considered above and still found wanting, in this modeled Texas ranch rangeland grazing induced added carbon uptake rates come nowhere near what it takes to negate the steep carbon costs of grass finished beef instead of permitting woody vegetation on it. If this is unambiguously so in both the upper Midwest[456] and semi-lush Texas[314]—definitely not true low productivity rangelands, but sites of active competition between grass finished beef and human food crops—there is little reason to expect such offsetting elsewhere, or indeed anywhere on a large scale.

Since even with all possible carbon sequestration credits the grass finishing operation can enjoy, the beef option still exacts carbon costs, not benefits, relative to the "reforest the cropland and rewild the rangeland" option, the latter is clearly the carbon favorable option.

This carbon favorability is all the more powerful given that we derived it under generous allowance for sequestration. For example, we can compare the 150 kg C (550 kg CO_{2eq}) per ha per year of added rangeland

AN ALTERNATIVE LOOK AT THE PUBLIC CONFUSION ABOUT THE ENVIRONMENTAL 81

sequestration due to grazing cattle to some soil organic carbon uptake rates I describe in the Accompanying PDF. Semi-arid grasslands managed under the USDA's Conservation Reserve Program took up 130 kg C per hectare per year,[343] a Colorado shortgrass steppe registered 20–40 kg carbon per hectare per year,[265] and a northern Plains site[673] ebbed and flowed widely following precipitation variability, yielding a long-term mean carbon uptake essentially indistinguishable from zero. And these are total carbon uptakes, not added uptake due to cattle grazing. While not impossible, therefore, carbon neutral grass finishing operations are the rare exception, definitely not the rule; most grass finishing beef operations are carbon sources.

Given the above results, let's see what difference *you* can make, assuming you, like the "average American," consume 6–6.5 kg beef protein a year. Forgoing this beef protein and favoring the "reforest the cropland and rewild the rangeland" option would save 0.1–1.1 metric ton CO_{2eq} a year, permit rewilding of 0.3–0.6 rangeland hectares, and allow reforestation of 0.1–0.2 cropland hectares. While these emission savings are definitely important, they are modest, 1–8% of our per capita total emissions (14 metric ton CO_{2eq} per person a year). The rangeland and cropland savings, on the other hand, amount to 40–75% and 25–50% of the total rangeland and cropland we use (about 0.8 and 0.4 grazed rangeland and harvested cropland hectares per person[57,626]). Since all costs of maintaining the beef grass finishing well exceed those of the "reforest the cropland and rewild the rangeland" option, that option is also the clear winner overall.

It is worth concluding this thought experiment by highlighting the great potential biodiversity benefits of the land savings of reforesting the cropland and rewilding the rangeland. One way to visualize these benefits is based on the degree to which every point on Earth's surface is currently impacted by human activity. But interpreting this information is made extremely difficult by great geographical variability in wildlife harboring potential (think, e.g., of the number and uniqueness of species that would have occupied your back yard if left alone for a century vs the same measures in an undisturbed Madagascar forest). Consequently, a particularly meaningful way to use this human impact measure at any given point is to normalize it by the average impact at all other points on Earth's with a similar wildlife harboring potential. By correcting for this great geographical variability, this puts all points on equal footing, giving a precise, uniform meaning to the comparison of any two points. Examining a map of these ratios[402] (their figure 4A) from the US Pacific coast to, roughly, Chicago, including both the rangeland and cropland portions of most mixed grass-finishing operations,[314,456] clearly shows[402] vast areas of very highly negative relative human impacts. And such areas of disproportionately high current human alteration, which forgoing beef will spare, are exactly the areas that can benefit from rewilding the most, because the life they now harbor is far below their biodiversity harboring potential. Clearly, rewilding rangelands and croplands that are now claimed by grass finishing beef operations stands to deliver considerable and important wildlife benefits.[157,197]

There is, of course, one key possible objection to this option: with lush cropland reforested and rangeland rewilded, under this alternative option, these spared lands, which now yield beef, would yield no food. This concern is addressed by the next option explored.

Alternative Option 2: Grow Some Alternative Food on the Lush Cropland, Rewild the Rangeland

A variant of the reforest and rewild option that addresses the "no food" objection is to rewild the full rangeland portion of the operation plus some of the cropland portion, but still keep in agricultural use enough of it to produce foods for direct human consumption that deliver as much protein as the forgone beef. Let's explore this option using a highly simplified version of our 2019 analysis.[183]

As alternative protein sources, let's choose both animal-based items—chicken, eggs—and plant-based ones—legumes (peas, lentils, or chickpeas) and cereals (wheat, rye, oats). For beef emissions, we will use the 60–170 kg CO_{2eq} per kg beef protein (operational emissions minus carbon sequestration by the grazed cropland portion of the operation) we derived earlier. Characteristic emissions of the animal- and plant-based alternative items are[475] 38–43 and 4–8 kg CO_{2eq} per kg protein. Every kilogram of beef protein replaced by a kilogram of alternative animal- or plant-protein thus saves operational emissions of 15–120 or 50–160 kg CO_{2eq} a year. For the average person who replaces beef with equal

AN ALTERNATIVE LOOK AT THE PUBLIC CONFUSION ABOUT THE ENVIRONMENTAL   83

protein amounts from other animal- or plant-based sources, this means saving operational emissions of about 100–850 or 310–1,100 kg CO_{2eq} a year.

Rewilding of 0.3–0.6 hectare rangeland per person choosing to replace beef with alternatives that exactly replenish the lost beef protein is, as in the preceding scenario, the key benefit to biodiversity and combating anthropogenic climate change. This spared area can be potentially enhanced by also sparing some cropland, which is made possible by the difference in cropland needs per person, 0.15–0.2 hectares per beef-eating person vs 0.01–0.04 hectares per person favoring the protein-conserving alternatives.[475] Every person who chooses to replace beef with alternatives would thus also save about 0.1–0.2 cropland hectares.

One possible use for this spared cropland is again reforestation. Taking the same conservative cropland reforestation carbon uptake we used earlier, 2,200–4,400 kg CO_{2eq} per hectare per year, the spared 0.1–0.2 hectares can sequester 250–880 kg CO_{2eq} per person per year, raising the total annual savings per person who replaces lush cropland–based grass-finished beef with animal- or plant-based alternatives while holding protein consumption fixed to 350–1,750 or 550–2,000 kg CO_{2eq} (some 3–14% of the US 14 metric tons a year total per capita emissions).

So even with reduced reforestation-related carbon uptake due to using some of the cropland portion of the operation to produce alternative food that maintains pre-replacement protein intake, the benefits of replacing grass-finished beef with animal or plant alternatives are almost unchanged, and all positive: averting emissions of roughly 0.4–1.7 or 0.6–2 metric tons of CO_{2eq} per person per year, rewilding 0.3–0.6 rangeland hectares, and sparing 0.1–0.2 cropland hectares per beef-replacing person that could be partitioned among additional rewilding and reforestation, and growing other crops. Let's get a sense for the potential national scale impact this transition may have. As a yardstick, let's consider the area currently allocated nationally to apples, oranges, vegetables, and oats, about 2.6 million hectares.[439]

How many grass finished beef eaters would need to replace beef to double the area available for producing these nutritionally essential staples? Dividing 2.6 million cropland hectares by 0.1–0.2 cropland hectares per beef-replacing person yields 13–26 million people, just 4–8% of the

U.S. population. This means that if a mere 4–8% of all Americans chose voluntarily to replace beef with protein conserving alternatives under this mixed option, we would double the area we now employ for the production of these food items, plus rewild 4–16 million hectares of rangeland (6–56 times the area of Yellowstone National Park), and avert annual emissions of 5–50 million metric ton $CO_{2}eq$. For comparison, one survey suggested that in Germany, plant-based protein accounted for 18% of total[391] consumed protein. A similar estimate in the US suggested that, assuming today's prices, as much as 28% of the population would favor one of the beef alternatives over beef itself.[616] Qualitatively and quantitatively consistent, even similar, results were recently obtained independently.[63] So such doubling is entirely possible.

By replacing grass finished beef with alternatives citizens could, on their own terms, reduce their emissions by 3–14%. While modest, this is significant. In recent years, world nations have striven, some more, most less, to reduce their carbon intensity by combining anemic legislative, executive, and private sector emission reduction efforts. In the US, for example, these efforts have borne some fruit, reducing emissions by about 15% per decade. So replacing beef with viable protein rich alternatives can save, simply and immediately, almost as much emissions as three to ten years' worth of an incredibly bureaucratically cumbersome, expensive, and politically fraught national project. Also, recall that we're talking about *all* emissions, including those associated with our needs for all agricultural products, electric power, housing, transportation, and myriad industrial outputs. No single silver bullet would address this diversity fully, keeping global mean surface temperatures no higher than 1.5°C above preindustrial levels, as the Paris Agreement stipulates.[578] Instead, many different solutions, each with modest benefits,[336] are necessary, which makes the beguilingly inconsequential 3–14% emission reduction essential.

Beef consumption is thus a unique environmental lever we can collectively pull on, or not. Beef contributes only about 10% of the protein we eat,[187] and our diet, in turn, is but a single dimension of our multifaceted resource intensity. Yet doing away with producing beef on lush grasslands in exchange for the many readily available, popular alternatives would guarantee reducing total US emissions by no less than 3%, and likely

10% or more. And this simple voluntary action, requiring no propriety or speculative technology nor Herculean legislative achievements, would also spare 38–75% of all US rangeland and 25–50% of its cropland.[57] This would dramatically reduce our appropriation of Earth's surface, allowing vast tracts of geographically diverse lands to regain their biodiversity potential. For many species, this may spell the difference between persistence and extinction. Beyond sheer magnitude, the savings calculated above are very robust, mostly based on readily quantifiable factors, and are *all* positive, even in the variant that conserves protein production but shifts its sources from beef to alternatives. Few other single changes match or eclipse any of these effects individually, and none that I can think of achieves all at once.

The resource savings we've found in this chapter are wholly inconsistent with the view of grass fed beef as environmentally beneficial. Because they address a beef variant that is likely the most carbon beneficial, those savings are also at odds with the view of beef in general as environmentally beneficial. With minute contributions to sustenance but major contributions to total carbon emissions, land appropriation, and other environmental burdens, grass fed beef is instead a costly food item.

6

Beef Production Using Presumably "Optimal" Grazing Practices

In chapters 4 and 5, we encountered preliminarily various grazing practices that some view as "optimal"[97] because they strive to best mimic the natural grazer–grassland dynamics perfected by evolution since the dawn of open grasslands.[206,515–517] These practices go by several names,[541] of which "holistic management" or "adaptive multi-paddock grazing" are most common, and stem from the presumed benefits of using brief, intermittent high stocking rate grazing. This is said to promote carbon sequestration in grazed lands enough to fully offset operational emissions and make beef environmentally sound, even desirable. Unfortunately, they are variably and often nonspecifically defined, and at least initially, these definitions have been frequently updated, especially in response to results that fail to reproduce the key claims.

I am deeply sympathetic to these views. When you witness the near-perfect, mutually beneficial interactions of grazing cattle and well-managed, thriving grassland, you cannot help but feel a deep sense of comfort that "all is well with the world after all." These sensations, along with my love of biking, have guided one of my favorite annual rituals. It takes place in December following the week-long American Geophysical Union's Fall Meeting in San Francisco, which I used to attend religiously as a graduate student and postdoc, and still do on occasion. After the meeting's final day, I'd go rent a road bike from a San Francisco bike shop, take the ferry to Larkspur Landing, and ride to the Point Reyes lighthouse. The last twenty or so miles, after leaving scary Route 1, mostly traverse lush grasslands dotted with healthy looking, robust grazing beef cattle. And this is a week before Christmas, when the Great Plains and most other Northern Hemisphere grasslands are mostly under a foot of snow! This scene of almost unparalleled beauty has been one of the highlights of my winter, documented with gigabytes of Point Reyes cattle photos accumulating on my phone, one even promoted to my desktop background. If there were only ten million of us to feed globally, I happily concede, such magical grasslands would never be used for anything other than full rewilding or ruminant grazing.

But there are eight going on ten billion of us. This quantitative difference calls for qualitatively different thinking. With so many of us, we must carefully consider all potential uses (including ones whose benefits may require a bit more subtlety than economic thinking offers) of every iota of resource, calculate all costs and benefits of every considered use, and favor the optimal use these calculations identify, the one that demonstrably maximizes our collective expansively defined well-being. And in this reality of limited resource availability, choosing to continue to use grasslands for ruminant grazing becomes more debatable. The focus of this chapter is this debate.

At their core, those practices advocate a deeply sensible grazer management that is based on dividing the total area of a given grazing ranch into multiple small paddocks through which the herd rotates. In each, brief, high stocking rates grazing sessions punctuate relatively long recovery periods that allow for adequate forage rest and recovery. This protocol is meant to mimic wild ruminant herd migration (such as bison or elk) driven by predation and grass availability. The practices also involve some human management elements that appear to me insufficiently reproducible and actionable.[570]

A very persuasive and ecologically sound motivation guiding these practices is overcoming grassland botanical deterioration by selective grazing. This addresses cattle's notorious "choosy" grazing style, the tendency to overgraze vegetation types they favor and shun other types they dislike, notably but not exclusively encroaching woody plants. This gives the unfavored vegetation types a competitive advantage over the favored vegetation—they themselves are left intact while their competitors are constantly clipped—raising the odds that disfavored plants will outcompete the desirable ones, surrendering the paddock to the terminal stage of grasslands in lush enough places: return to a woody community where grazing is energetically unrewarding, hard, or altogether impossible.

These ideas are ecologically reasonable, and indeed verifiably materialize in some settings (e.g., in two ranches in lush, topographically accommodating southern Michigan[49]). Consistent with the above motivation, these approaches appear to benefit not only botanically desirable grassland attributes but also richness and diversity of soil microbial communities.[311] It also appears that they may underperform in the harsh realities of commercial grazing, delivering inferior grass digestibility and protein content[292] and weight gains inferior relative to conventional continuous grazing.[146]

Some Background

Holistic management was first introduced to general audiences in a widely viewed TED talk[515] by the method's original spokesperson, Allan Savory. In that talk, Savory mostly shuns specific, falsifiable predictions, favoring instead such extravagant claims as "green the desert!," "reverse climate change!!,"[515] and the like. Because they sounded too good to be true, the claims, and the talk, quickly attracted wide public and professional attention, and were vigorously contested.[97,434] Notably, they piqued the curiosity of a group led by Texas A&M botanist/range scientist David D. Briske. Carefully examining Savory's key claims, the group rejected most[71,73] using well referenced, cogently argued evidence.

This could have been the end of this, but for the rejection of the critique[73] by a fellow Texas A&M scientist, Richard Teague (cited extensively in chapters 4 and 5 and the Accompanying PDF). After characterizing the Briske et al. critique as betraying "a very poor understanding of Holistic Planned Grazing,"[588] Teague highlighted what he viewed as the critique's Achilles' heel, reliance on "poorly executed grazing experiments."[588] This quickly solidified into holistic management's key defense, a sweeping rejection of results inconsistent with its ever adjusting basic tenets as experimentally flawed.

Unsurprisingly, this persuaded few. A particularly illuminating example of unpersuaded topical experts [116] addressed a key element in Teague's critique, which he viewed as epitomizing "[t]he folly of Dr. Briske's statement."[73,588] It focused on ranches in arid Patagonian Argentina and Chile that have been continuously grazed for several decades (in the holistic management worldview, continuous grazing is a key promoter of rangeland deterioration). Teague characterized these rangelands as in a bad and steadily deteriorating state despite low and declining stocking rates, which was central to his critique because it demonstrated perfectly the ruinous effect of continuous grazing. Or so he thought. The group that countered Teague's rejection [116] of Briske's critique [73] comprised South American rangeland researchers with decades of local experience between them and intimate familiarity with the rangelands at the heart of this exchange. Synthesizing their rich rigorously observational perspectives, published papers,[84,442,443] and firsthand knowledge, they drew a diametrically opposite picture of these rangelands, pointing out that Teague provided no reference or novel published evidence about these lands. In one site,[442] they reported vegetation holding steady over a decade of continuous grazing, even in drought years. Similar findings—roughly unchanging total vegetation or forage species cover—also characterized another, drier site that was moderately continuously grazed for over twenty years,[84] and another where, importantly, sheep production indicators also didn't change over the follow-up period.[443] Overall, Teague's evidence was portrayed as at odds with observations by resident researchers deeply familiar with the relevant lands.

Some tenets of holistic management, as imperfectly as they can be gleaned from Savory's vague manifestos,[515,517] are incredibly naive about Earth's workings. For example, when Savory sets out to green Earth's deserts by cattle grazing,[515] he is also (unknowingly, I assume) declaring war on the stifling hegemony of angular momentum conservation and the geography of Earth's major mountain ranges, because it is these two variables that actually determine the distribution of Earth's deserts far more powerfully than effects of cattle.[103,351] Not only is this view of deserts dynamically naive, but because neither of these governing controls responds meaningfully to our decisions, it is also futile.

Other elements of holistic management views are mechanistically possible but not yet convincingly reproduced by unaligned researchers. This is a key reason for the lukewarm reception of these views by rangeland scientists (e.g., Briske et al.[72], Henderson et al.,[258], or Hawkins et al.[251]). But neither are these views ever emphatically debunked, because they are defined so broadly and imprecisely,[517] and are so varied in implementation[67,542] as to render them non-falsifiable. Systematic peer-reviewed analyses of the literature that examines the claims of holistic management[73,251,434] reveal a mixed record that, rather than corroborating the sequestration claims, raises further questions about them. A recent example[312] that analyzed twenty-four ranches in Alberta, Canada, with annual precipitation ranging roughly over 330–550 mm is a case in point. While soil organic carbon concentrations and stocks (given in grams of soil carbon per kg dry soil and metric ton per ha respectively) were modestly higher on average in the adaptive multi-paddock systems as compared to conventional ones, considerable variability rendered either difference insignificant. So in sites more or less ideally suited for adaptive multi-paddock systems in terms of topography and climate, the practice proved indistinguishable from conventional continuous grazing. Other differences—such as in the association of resident organic carbon with fine particles, or in carbon to nitrogen ratio—were significant. These are intriguing, and potentially important and actionable if proven robust and reproducible by further studies, but currently tentative. It is possible that these changes will become mechanistically attributable to systematic restructuring of soil microbiota by adaptive multi-paddock grazing,[311] but this too is currently too insufficiently vetted.

Some confirmatory publications on positive effects on some grassland indicators exist, some methodologically sound.[102,193,634] Other either lack quantitative specificity[589] or report soil attributes under holistic management not materially different from those in grazing exclosures.[501]

None of the above challenges outright negates the possibility and tentative evidence that various well-managed rotational grazing approaches at stocking rates consistent with local carrying capacity and primary productivity can modestly improve livestock productivity and soil carbon stocks compared to continuous grazing in such temperate locales as most of the US and Europe.

But the key trouble with many of the papers that conform with the key tenets of holistic management (e.g., Kim et al.[413] and Mosier et al.[314]) arises from the lands they analyze. Surely not arid, often not even semi-arid, in terms of annual amount and seasonal distribution of precipitation, and mostly topographically accommodating, these lands are for the most part croplands, not rangelands (see Accompanying PDF). As such, they are expected to support aboveground productivity and sequestration rates well above what can be reasonably expected of actual rangelands, where cattle grazing does not compete with or displace alternative forms of food production. Focusing instead on lush croplands[314,413] means disregarding direct competition between grazing ruminants and not only wildlife or reforestation (that in most locales stand to reproducibly enhance carbon storage far more than grazing) but also production of food for direct human consumption that can deliver far more protein and improve nutrition at far lower resource costs, as we have already seen aplenty. Disregarding these relevant competitions means that the levels at which Mosier et al.[413] or Kim et al.[314] set the bar cattle grazing must clear in order to be declared environmentally desirable are so artificially low as to have no clear environmental meaning.

Soil Conservation Concerns

There is a potentially fatal flaw in my critique above, stemming from soil conservation concerns. It arises because many areas that are lush enough for crops yet are used for grazing today have sandy soils that make them susceptible to soil erosion if tilled.[540] Yet, while most grassland vegetation is typically perennial, most croplands are tilled. If repurposed for growing alternative food, therefore, currently infrequently tilled grazed croplands too will likely be tilled, raising the prospect of land degradation. This suggests that their only safe use is perennial grass for grazing. In repurposing them to cropping anyway, then, are we not committing the same environmental folly that wrought the Dust Bowl calamity and hollowed the Plains of much of their population and economic backbone in the 1930s,[166,654] and that brought once great empires to their knees in our more distant past?[406]

This is a sensible question. In a perfect world, we'd exempt such lands from any use other than rewilding or conservation-minded, light-touch grazing.[346,408] But because we must eat, is there a way to use such lands responsibly, maintaining the requisite unwavering commitment to land and soil conservation?

I think there is, because a hectare yields so much more when allocated to legumes or cereals for direct human consumption than when allocated to beef. We can thus easily allocate no more than 10–15% of any parcel of land for such crops and still produce enough protein to more than offset the loss of beef protein. The key is to combine this admittedly imperfect environmental compromise with an ironclad commitment to soil conservation that relies on aggressive deployment of such demonstrably successful soil conservation measures as periodic fallowing,[2] replacing annual plants with hardier perennials,[528,674] and intercropping,[422] among others.[295,558] For example, in the scenario we analyzed in chapter 4, using only 15% of every cropland hectare for alternative crops still yielded 2–3 times as much protein as would beef using the full hectare, with the remaining untapped 0.85 hectare allowing more than enough farmer discretion and flexibility for strict soil conservation measures. Consequently, despite the potentially reasonable soil conservation–based objection, lands such as those Mosier et al.[413] or Kim et al.[314] analyzed are croplands on which ruminant grazing directly inefficiently competes with production of alternative human foods.

A Final Take on "Optimal" Grazing Ideas

Holistic Management

With my view of the soil conservation issue stated, we turn our attention back to the question of whether adaptive multi-paddock grazing outperforms conventional continuous grazing. The question arises mostly because proponents of adaptive multi-paddock grazing often dismiss such discouraging results as surveyed earlier as reflecting "the wrong type of grazing,"[208,413] invoking such supportive results as those of Mosier et al. 2021.[413] Of course there remains the issue we encountered earlier, that many of these results are obtained on croplands easily usable for most forms of human food production, not actual rangeland. This notwithstanding, some solid papers suggest that multi-paddock grazing outperforms conventional continuous grazing,[314,413] while others are mixed.[518]

Many holistic management studies are suggestive yet hard to draw firm conclusions from. A good example [15] compared various ecosystem functioning measures (including vegetation biomass and composition, water infiltration rates, and soil carbon storage) in five adaptive multi-paddock ranches in the southeastern US to those of five neighboring ones that used traditional continuous grazing. Let's not belabor the point that these are, once more, all essentially croplands. In three out of the five pairs, the adaptive multi-paddock ranch had more, in some cases *much* more soil organic carbon than the neighboring continuously grazed ranch. On the face of it, this looks like a solid ecological and carbon-related observational vindication of holistic management. But on closer examination, the water gets a lot murkier. First, while the authors went to admirable length to match pairs of like ranches, the likeness permitted by the available ranches is imperfect at best (table 1 of Apfelbaum et al. [15]), suggesting that some of the observed differences arose from comparing likes to unlikes. In addition, the duration under each of the two current management practices in each of the five pairs differed by a factor of 2–4, likely placing the ranches compared in each of the pairs at very different stages of evolution from their unique initial states to their individual carbon saturated states. How do we know that the observed differences are not due to one system having not yet caught up with the system whose deployment is longer? While the authors mention data collection also in neighboring reference natural areas, if such data exist, they are excluded from the paper, [15] another missed opportunity to investigate the origins of the observed soil carbon differences. Most troubling of all to me is the absence of the soil carbon content of the ranches at the time the current grazing practice was first implemented. Without this initial condition, and with the unequal duration and unmatching histories of the two ranches in each pair, the observed higher soil carbon content under holistic grazing cannot be attributed to this grazing method, because it can simply reflect more favorable initial carbon contents, or different stages of carbon saturation. These likely possibilities are neither mentioned nor analyzed. These limitations all but preclude determination of mean annual carbon uptake rates, let alone temporal evolution of these rates as soil carbon saturation nears. Consequently, the very interesting carbon excess Apfelbaum et al. [15] report may hint at carbon uptake enhancement by holistic grazing, but less conjectural plausible alternative explanations are neither investigated nor ruled out, mostly because they *cannot* be ruled out. Enhanced carbon uptake by holistic grazing thus by no means straightforwardly follows from the findings of Apfelbaum et al.[15]

Doubts about the supposed superiority of holistic grazing deepen with a recent review that focuses on Queensland, Australia, but uses global literature.[259] Mirroring my own assessment of Apfelbaum et al.[15] above, this review found it difficult to form representative statistics from reported sequestration rates because of methodologically inconsistent trials and surveys; inadequate "control" or "baseline" sites; different analytic methods for soil organic carbon evaluation; and inadequate account of the effects of variability in time and space of climate, soil type and composition, landscape, and vegetation that all control carbon uptake much more strongly than different management practices. Despite the din, there was enough of a signal for the review to find that, in Queensland, as in what was covered in the last two chapters, high stocking rates of holistic management reduce soil organic carbon, that reducing stocking rates or excluding grazers altogether minimally increases soil organic carbon, and that soil organic carbon contents under rotational and reasonably stocked traditional continuous grazing are indistinguishable.

Because reforestation is a key alternative land use contender in the considered scenario, the following finding is particularly important: converting woodland to perennial grassland reduced soil carbon uptake by 620 kg per hectare per year in some places, and increased it by 120 kg per hectare per year in others, yielding a highly uncertain mean carbon uptake decrease of 190 kg per hectare per year. In other words, higher carbon sequestration when the land is forested, and less when it is grazed. It follows that in Queensland and likely elsewhere, in the scenario of switching from beef grass finishing to rewilding of all rangeland plus 85% of the cropland, reforesting the 85% of the land that is not used for alternative crop production in the grazing cessation scenario would at the very least not constitute a carbon cost, and would quite likely confer substantial carbon benefits.

A similar story is told by methodologically sound, careful performance comparisons of adaptive multi-paddock rotational management to traditional season-long continuous grazing in northern Colorado rangelands.[24,25] The setting is key: an almost arid, actual shortgrass steppe rangeland, where competition with alternative human food production is minimal. While the authors' focus was not carbon sequestration, they did evaluate a key factor impacting the operation's overall carbon budget, cattle weight gain rates. In six years, these rates were on average 15% lower under the adaptive multi-paddock management than under continuous grazing. In the seventh year, in response to a drought and the low weight gain of the previous six years, the number of cattle kept in the adaptive multi-paddock system was halved, which led to equalizing weight gains. Lower beef productivity under adaptive multi-paddock rotational management was also later observed in the same geography, with season-long moderate grazing outperforming adaptive multi-paddock management in all but one of nine comparisons and yielding on average 10% more beef per hectare (table 1 of Raynor et al.[482]). With the high similarity of all compared paddocks, the 10–15% less beef per hectare under adaptive multi-paddock management yields operational emissions per kilogram of beef produced in this system being very robustly 11–18% higher than those of the moderately stocked, continuous-grazing system. Since as we have seen, sequestration rates are not reproducibly higher, and are quite often lower under holistic management, roughly the same is also expected to hold for net emissions (emissions incurred during the production process minus sequestration realized on occupied land). In other words, adaptive multi-paddock rotational management showed no superiority over traditional season-long continuous grazing with commonsensical stocking rates, and likely the reverse.

Beyond—or perhaps *because* of—repeated failures to observationally corroborate the supposed superiority of holistic management, its promoters sometime indulge in such philosophical sounding declarations as "There is no absolute truth in science" or "There is no such thing as an 'unbiased observer' in rangeland science."[590] Yet even in our so-called "post truth" times,[341] such pronouncements are useless as guidelines for such practical questions as whether eating beef is environmentally desirable; whether, if it is, grass feeding is superior to industrial feeding; or, if it is, whether holistic management is superior to light stocking continuous grazing. With the paramount environmental importance of these questions, which are fundamentally about how to best manage about a third of Earth's land surface, this pseudo-agnostic attitude becomes reckless. Imagine if these declarations did govern rangeland science; why bother with measurements then? And if we shouldn't, thus defanging rangeland science of quantitative rigor, how is the resultant pursuit any better than theology? The absence of conclusive evidence that adaptive multi-paddock rotational management outperforms reasonably stocked traditional continuous grazing in grass or beef yield, carbon sequestration, or ranch profit, and the above logical failures thus cast holistic management as an idea with some intuitively appealing elements and some questionable ones, offering some intriguing possibilities but few firm, currently actionable contributions to the discussion on environmentally sound beef.

Other Suggested Beef Grazing Practices

Let's next briefly consider regenerative grazing,[225,328] whose objectives partly overlap with holistic management's. Scientifically and conceptually, the practice is extremely compelling, potentially enhancing vegetation productivity, soil health, resilience to climate change, health and nutrition of grazing animals, erosion control, and ranch profitability and financial resilience.[565] But whether this potential can be realized on a wide enough scale to make a real difference remains to be seen. One comparison of regenerative and conventional rotational sheep grazing[140] revealed 30% higher springtime grass production and 4% higher topsoil carbon storage under regenerative than under conventional rotational grazing. Another such comparison[140] was disappointing, showing (figures 1–5 of Otalora et al.[140]) that for such critical ecosystem measures as springtime grass production, topsoil carbon storage, water flow regulation (the soil's ability to retain precipitated water), activity of enzymes central to nutrient cycling rates, and microbial biodiversity, variability within each treatment plot proved large enough to render the two treatments essentially indistinguishable. Only two of fourteen tested differences between the two practices—14%—proved statistically significant (rightmost column of table 1 of Otalora et al.[140]). Such results tentatively suggest that regenerative grazing may outperform conventional grazing sometimes, but as of summer 2024, solid information is insufficient for firm conclusions.

Next, we turn to "mob grazing," extremely brief transient grazing at exceptionally high transient animal densities. A careful eight-year comparison of this to other management options in a Nebraska Sandhills meadow[547] revealed some real differences. Mob grazing cattle consumed markedly less available herbage than in other grazing practices, but trampled without consuming much more, leaving far less standing dead vegetation. All grazed plots had mutually indistinguishable dead and litter biomass that were markedly below those in ungrazed control plots. But, consistent with the observation of eating significantly less of the available forage and as in the Augustine et al. results reported earlier,[24,25] something else was distinct about the mob-grazed plots: steers grazing them gained much less weight, about 150 g per day compared with 280–645 g in other grazing treatments. While we need to know more about mob grazing, the results thus far are underwhelming.

The asserted superiority of proposed "optimal grazing" alternatives—including, notably, holistic management, adaptive multi-paddock grazing, mob grazing, and regenerative grazing—has therefore not been demonstrated. This conclusion is strong in croplands used for grazing, and stronger still in true rangelands. The notion that holistic management or its peer practices would render beef environmentally sound by offsetting its meticulously documented exceptionally high operational emissions is thus unwarranted and overly conjectural. Negating very strong evidence requires extraordinarily strong counter evidence, which the "superior grazing practice" literature comes nowhere near. With our current imperfect knowledge, the prospect that these alternative practices would eventually become ubiquitous is thus deeply troubling, because such resource allocation would very likely prove suboptimal, failing to maximize societal benefits while needlessly enhancing environmental impacts.

II

Fundamentals

7

A Brief Cosmic–Planetary Background

One of the most common agricultural practices, familiar to even the most casual observers of agriculture, is fertilization, the addition of nutrients—mostly nitrogen, phosphorus, and potassium—to enhance productivity. Starting with the most basic level, the reason these nutrients are agriculturally necessary is their key roles in life in general.

Nitrogen is crucial because much of the biochemistry of life is catalyzed by enzymes, regulated by hormones, and carried out by muscles, all peptides or proteins. Since peptides and proteins are indispensable components of life, and are all polymer chains of amino acid molecules whose biochemical functionality arises from a perfectly placed nitrogen plus hydrogen group, nitrogen is essential for life. Life also requires phosphorus because phosphate, a phosphorus atom bonded with four oxygen atoms, is central to biological energy metabolism and transfer, and to DNA structure, among other biological uses. Potassium is likewise life essential because it plays key roles in maintaining electrochemical gradients across membranes, as essential to life as can be. Additionally, in multicellular organisms, potassium, along with other such singly charged ions as sodium or chloride, helps regulate blood pressure, acid–base balance, and conduction of nerve impulses.

Life thus critically depends on nitrogen, phosphorus, and potassium, on carbon as CO_2, and on hydrogen and oxygen as water, and life's churning—the constant synthetic building and imperfect metabolic recycling of enzymes, hormones, and muscles—requires a steady supply of these nutrients. Since agriculture is all about enhancing productivity by accelerating life's pace, it depends even more critically on them. Meeting these requirements is what spreading of manure or synthetic fertilizer achieves, explaining its centrality to agriculture.

Why life evolved around the above light elements is a heady question we will not negotiate seriously here. ("Light" refers to atomic mass of an element, e.g., 4 and 16 for the prevalent isotopes of helium and oxygen, the number of protons plus neutrons in its nucleus.) But some basic connections are inevitable. First, life simply had to make do mostly with abundant stuff; it is hard to imagine a life form that depends critically on macroscopic amounts of uranium, say, thriving and taking widespread hold. Indeed, the above life-essential light elements are, to variable degrees, relatively abundant in the bulk earth, the solar system, and beyond. (But what matters most to agriculture are the Earth crust and atmosphere abundances into which Earth physics, biology, and biogeochemistry translated "background" abundances.)

The key relevant chemical attribute of the above light life-essential elements is high reactivity (as measured by electronegativity, the propensity to attract shared electrons). For example, water, the universal solvent and the medium in which known life unfolds, owes many of its key properties (including the "universal solvent" title) to its polarity, with electrons distinctly favoring the very electronegative oxygen over the less electronegative hydrogens (which arises from these atoms' number and configuration of electrons).

Another, related basic attribute that endows carbon, nitrogen, and phosphorus with the chemical flexibility life requires is their wide range of oxidation states, how readily they donate or accept electrons when forming molecular bonds. This means that they can play widely varied roles in a suite of chemical reactions—redox chemistry—particularly important to life, electron exchanges between individual atoms in a molecule. For example, carbon can accept four electrons from four hydrogens, forming methane and attaining a −4 oxidation state (it accepted four electrons along with their four negative unit charges), the most electron rich it can be. On the opposite extreme, in CO_2, it surrenders two electrons apiece to two hyper-electronegative oxygens, attaining a +4 oxidation state (it now donated four electrons, losing four negative unit charges). Carbon's flexibility for being anywhere between these extremes—notably as the key element in organic matter, with an oxidation state of zero—and the rich spectrum of organic and inorganic chemistry it can prominently participate in are key reasons the carbon cycle is as complex as it is, spanning the atmosphere, biosphere, soil, upper ocean, deep ocean, sedimentary rocks, mantle, and crust, and spanning timescales from seconds or less to hundreds of millions of years, some fifteen orders of magnitude. Similar considerations also apply to nitrogen, but the oxidation state range is shifted one notch up (i.e., spanning +5 to −3) because of nitrogen's one extra electron relative to carbon.

These chemical elements, and most life's driving force, solar energy, have tightly entangled biographies.[253,332,507] They all arise from a process we encountered briefly in the Prologue, the origin of the chemical elements in general and the light ones relevant to agriculture in particular. Let's revisit this question in a bit more detail.

The Cosmochemical Background

Recall that in the early universe right after its Big Bang birth, all ordinary (not dark) matter comprised mostly hydrogen, some (around 25%) helium (the lightest two elements, with dominant atomic masses of 1 and 4), and very little else. Additional helium, and heavier elements, including all the ones that participate in life and agriculture, have been created since then by nuclear processes in stars. Really heavy elements—in general tangential to agriculture—are synthesized in high energy routes associated with the final acts of large-to-massive stars (double and tens of times the solar mass, respectively) we will not discuss here.

Conversely, elements with atomic masses 7–40, including the agriculturally central ones mentioned above, are of great interest to us. They are nucleosynthetically produced, the daughters of lighter parent elements that underwent nuclear fusion in cores of older stars.

As a preamble, do not confuse the following story, creation of atomic nuclei from lighter ones by fusion, with chemistry, which is about electrons. In stellar interiors, energies are far too high for matter to exist as whole atoms, with a full set of electrons neatly swarming around specific nuclei. Instead, fusion addresses nuclei, where most mass and all positive charges of atoms reside. Because chemistry involves electrostatically neutral whole atoms, with as many swarming electrons as protons in their nuclei, it unfolds at vastly lower energies (temperatures), following very distinct physics.

Because protons are positively charged, they mutually electrostatically repel when near one another, with the repulsive force quadrupling for halving the distance. Left to their own devices, therefore, neighboring protons would remain solitary. If they are somehow compelled to come very near each other (but not too near)—hurling toward each other at high speeds characteristic of very high temperatures is one way for doing so—an attractive force, the nuclear (or residual strong) force, becomes dominant over electrostatic repulsion. In the range of distances between nucleons (protons or neutrons) in smaller atomic nuclei—of order a fm, comparable to the radius of the helium-4 nucleus (where fm means femtometer, a million-billion-th of a meter), attraction due to the nuclear force is 10–100 times stronger than electrostatic repulsion. This makes binding protons and neutrons into larger nuclei energetically favorable, *provided* the force necessary to overcome electrostatic repulsion exists. (A process is "energetically favorable" if its unfolding releases energy.)

The above situation in the nucleus can be likened to being tasked with moving a rock from a given plane you are on to a topographically lower plane, but with a commanding hill ("potential barrier") between the two planes. Pushing the rock against gravity to the summit is the "energy costing" part of the journey, standing for the work required to bring two protons closer together against electrostatic repulsion. But once just past the summit, letting go means the rock spontaneously rolls down toward the lower plane, performing *more* work against rolling resistance than the work invested in bringing it to the summit. Continuing with the two-proton scenario—and assuming sufficient forces were applied on them against mutual repulsion to bring them close enough together for the nuclear force to match and then exceed electrostatic repulsion—the spontaneous rolling part of the rock's journey is an apt metaphor for the two protons "falling" toward each other under the attractive nuclear force, attaining a more stable configuration than they had when unassociated, releasing the energy difference to their surroundings.

The above is quite consistent with our macroscopic world intuition. But the next pivotal element of the story—how the energy difference is released—highlights that very small scales are not miniatures of our familiar macroscopic world. For the rock, it is straightforward: available gravitational potential energy (arising from the hilltop sitting above the destination one) transforms into heat the rock generates as it rolls across the landscape, or into work performed on destroyed objects in its path. But for the fusing protons, this intuitive simplicity is lost.

Luckily, this nonintuitive energy release is governed by one of physics' most celebrated equations, Einstein's mass–energy equivalence. This is most simply illustrated with the fusion of a neutron and a proton into the nucleus of one of hydrogen's isotopes, deuterium. In the course of this fusion, a photon—the carrier of electromagnetic radiation—is emitted. But photons carry energy. The photon loss thus implies that the fusing nucleons lose energy, attaining as the occupants of a deuterium nucleus an energy level lower than their combined standalone energies. That energy loss, call it ΔE, is proportional to a corresponding mass loss, or "mass defect," Δm, by the famous relation

$$\Delta E = \Delta m\, c^2.$$

Since c is the speed of light, approximately 300 million meters per second, the proportionality constant c^2 is rather colossal, almost 10^{17} m^2 s^{-2} (1 with 17 zeros to its right). Let's apply this relation to the fusion of protons into helium-4 nucleus (two protons and two neutrons). The mass defect of helium-4 is about 0.7% of the sum of masses of its four constituents. If all parent masses are converted into daughter masses, the fusion of a kilogram of protons into helium-4 is expected to yield a kilogram of helium-4. Instead, we get about 993 g, with the remaining 7 g releasing

$$\Delta E = \Delta m\, c^2 \approx (7\,\text{g}) \cdot (9 \times 10^{16}\,\text{m}^2\,\text{s}^{-2}) \approx 6 \times 10^{14}\,\text{joule}$$

of energy; seven grams that fully meet the energy needs of New York City for about thirty-two hours!

These general principles extend beyond the above simplest and most energy-dense fusion process. Provided high enough energies are present to overcome electrostatic repulsion, which rises as nuclei get bigger and hold more positively charged protons, existing nuclei heavier than a proton can further fuse into heavier nuclei still. But even with enough energy, there is an upper bound on this process. Recall that the binding of nuclei by the nuclear force is most effective over very short distances, but falls off rapidly with increasing nucleus size as nucleons of opposite sides of the nucleus recede ever further from one another (e.g., the nuclear attraction between such nucleons is roughly a thousandfold weaker in lead-208 than in helium-4, for a corresponding twelvefold diameter increase [1,393]). Because very large nuclei are therefore less tightly bound than nearby lighter ones, some of them can yield energy not through becoming larger still by permanently fusing additional nucleons, but by going in the opposite direction. That direction, fission, describes larger nuclei splitting into smaller ones and emitting energy, either spontaneously very slowly or (as in nuclear power plants) after being destabilized by neutron absorption. This too is of no further interest to us here, but what is important is the transition between the regimes, which marks the upper mass bound on fusion, the set of most tightly bound nuclei in the iron–nickel–copper group (mass numbers 56–65).

This duality—in which energy release accompanies either light nuclei fusing into heavier ones or very heavy ones splitting into lighter ones— is illustrated by figure 7.1, which presents nuclear binding energy per nucleon (proton and neutron) as a function of mass number (size of the nucleus). It is shown schematically here, with the vertical axis being a measure of the energy per held nucleon it would take to undo fusion by breaking asunder a given whole nucleus back to protons.

The stability maximum introduced earlier (mass numbers 56–65, shown by gray shading) marks the heaviest nuclei the fusion regime on the left can yield, and separates it from the fission one on the right. Note the incredibly steep rise of the fusion regime between lone protons (the lower left end of the curve) and helium-4, whose binding energy per nucleon is roughly three- and sevenfold higher than its lighter neighbors deuterium and helium-3. It means that of the full energy gain the fusion regime permits—from protons to the iron group peak—almost 80% is already realized by fusing protons into helium-4. I visualize this as a 60 km bike ride (standing for the mass difference between a proton and

A BRIEF COSMIC–PLANETARY BACKGROUND

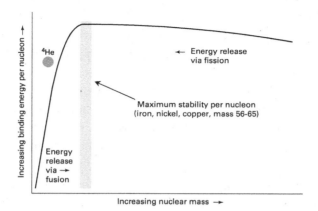

7.1 A schematic representation of the dependence of nuclear binding energy per nucleon (proton or neutron in the nucleus) on nuclear mass. The maximum stability per nucleon region (masses 56–65, including nuclei of iron, nickel, copper) is highlighted by gray shading. The helium-4 nucleus is also shown. Its position, well above the general climb of binding energy with nuclear mass between hydrogen and the maximum stability region, indicates exceptional stability. The maximum stability region (gray shading) separates distinct energy extraction regimes. To the left of the shading (mass range 1–56), a great deal of energy liberated by fusing light nuclei into heavier ones. Conversely, to the right of this region (masses above about 60), smaller amounts of energy are liberated by splitting heavier nuclei into lighter ones.

the iron group) up a hill about 880 meters high (standing for the binding energy difference between the iron group peak and solitary protons), with the initial climb so steep that after only 3 km (standing for the proton–helium-4 mass difference), a mere 5% of the way, you have already gained almost 710 m of elevation. This is how extraordinarily stable helium-4 is, and the reason it is disproportionately available in stellar cores as fodder for further production of heavier daughter nuclei.

These considerations play out in stars as they go through their life cycles. Stars arise in vast cold, dense molecular gas (mostly H_2) and dust clouds in such disk like galaxies as our Milky Way. Either randomly (purely by happenstance), or in response to external coherent forcing (such as a supernova explosion nearby), parts of the cloud can become denser than their surroundings, thus becoming gravitationally more attractive. Under this attraction, stuff from the surrounding disk accretes onto those denser spots, making them even more massive and gravitationally attractive. If enough matter is present in the vicinity of such an anomalously dense locale, and the competition for accreting matter with other nearby higher density spots is not too stiff, enough mass from the surrounding cloud can accrete locally to form small grains, pebbles, and eventually proto-stars. As protostars gain mass by "vacuuming" their galactic environment, the gravitational energy of the accreting matter and its compression by gravity into an ever-shrinking volume (which compresses the internal energy associated with random particle motions into this smaller volume) steadily raise core temperature and pressure. At some point, core temperatures become high enough (of order 10 million K), and the mean distance between neighboring protons becomes small enough to "ignite" a nuclear furnace in which protons fuse, mostly forming helium-4 and liberating tremendous amounts of energy, as described above. (Note that "ignite" is a metaphor, not a literal description because, recall, there is no chemistry here, just nuclear interactions. But "ignition" and "burning" are apt pithy metaphors.)

But how does the above fusion process generalize to heavier elements? How does proton "burning" climb up the mass ladder all the way to the iron group?

The first rung is climbed when dwindling hydrogen supplies suppress core proton collision rates below those required for continued hydrogen "burning." This puts the star's power generation on hold, because while at this point its core is basically helium, it is not hot enough for helium fusion (because two neighboring helium-4 nuclei, each with two positive charges, repel each other electrostatically more than two protons with one positive charge apiece, so higher energies are required to overcome the greater electrostatic repulsion). The absence of a central heat source changes the star's force balance. As long as hydrogen fusion continued, inward gravitational attraction toward the core was balanced by declining pressure away from the core maximum. With fusion discontinued, the pressure gradient gradually fades as the core cools, and gravity—now unopposed—shrinks the helium core. But, like in home heat pumps, compressing matter into smaller volume in the shrinking helium core reverses the cooling, steadily heating the core. Once core temperature reaches roughly a hundred million Kelvin, fusion resumes, this time of helium-4 into carbon-12, nitrogen-14, and oxygen-16. This produces the two most abundant (and life- and agriculture-essential) elements heavier than helium,[19,358] carbon and oxygen.

But not nitrogen, which is so central to life and agriculture. Even though it is an intermediate product in hydrogen fusion, nitrogen-14, midway between carbon and oxygen and *not* an integer helium-4 daughter, is less stable than carbon-12 despite being heavier. Heuristically explained in terms of having odd numbers, 7, of both neutrons and protons,[247] this yields[211] galactic and solar system nitrogen-14 abundance roughly an order of magnitude lower than those of carbon and oxygen, partly because it is fodder for their subsequent synthesis. This is important background to how nitrogen availability assumed such centrality to agricultural productivity.[211,358,652]

For massive enough stars, fusion can progress up higher rungs of the elemental mass ladder in "advanced burning" of carbon, neon, oxygen, and silicon, synthesizing heavier and heavier nuclei from the products of earlier burning phases. While this is happening in the core, lower mass burning characteristic of earlier phases can continue in the outer, cooler layers of the star, giving rise to an onion-like layered structure, with distinct concentric shells of dominant elemental compositions and fusion reactions. This culminates with silicon-to-iron and nickel burning in the star's core, endowing the elderly star with a core made of iron group elements. (While this is superficially reminiscent of Earth's iron and nickel core, the vastly larger stellar version arises from physically totally distinct processes, because Earth never had a fusing core.)

Eventually, even the largest star runs out of nuclear fuel and goes cold. This is accompanied by ejecting some or all of the matter it is made of, including the heavier elements it synthesized over its lifespans, into the interstellar medium. Consequently, as the universe ages, it gets enriched in elements heavier than helium-4 (its "metallicity" rises, in cosmochemistry and astrophysics parlance). Some of this enriched matter ends up in later generation accretion centers, where new stars, like our Sun, are built from slightly heavier initial compositions than those of their forebearers.

The cosmochemical processes described above sculpted the composition of the solar nebula, thus laying the foundations for the elemental abundances with which Earth's life has had to contend.[242,623] These abundances exhibit two clear patterns. First, abundance generally declines with mass, reflecting additional, more involved fusion processes required for assembling heavier nuclei. Second, abundances fluctuate, with elements with odd number of protons being less abundant than their even numbered neighbors (which applies to the anomalously low abundance of nitrogen, as we have seen earlier, between the two neighboring, also agriculturally critical and more abundant elements, phosphorus and potassium).

Abundance Modifications by Planetary Processes

From the solar nebula composition, individual planets within the solar system were further compositionally modified. A most important component of these modifications is loss to space of volatiles, low boiling temperature substances such as hydrogen, helium, carbon, nitrogen (which volatilizes above about $-150°C[-240°F]$ at any relevant pressure[205]), oxygen, or water. Such losses are substance specific (because of variable volatilities), and planet specific, depending on a planet's temperature, and on its mass and radius, which jointly determine its escape velocity, how fast an object must fly away from a planet to escape its gravitational grip. This reflects the nature of Newtonian gravity: the more massive a planet is, the stronger its gravitational pull, and the smaller its radius, the nearer enveloping particles are to the planet's center of gravity, with both reducing the planet's loss of these particles.

With its current mass, Earth has no problem holding on gravitationally to most all of its carbon, nitrogen, or water, as we will exemplify quantitatively shortly. But the small masses of primordial parent bodies like the embryonic protoplanets that eventually accreted onto Earth mean that atmospheric gas collected around each of them individually needed only to attain fairly modest temperatures (energies) to irreversibly escape into the surrounding interplanetary space.[237] For example, while Earth's escape velocity is just over 11 km per second, those of objects with Earth's density and 10 and 100 km radii are roughly 0.02 and 0.18 km per second, only 0.1–2% of Earth's. Some of nitrogen dominance over fertilizers thus arises because Earth is the final accreted body of many smaller objects, from which losing volatiles at a given temperature was 98–99% easier than it is from the mature Earth.

The surface temperature dependence of volatile loss rates reflects particle energy; the higher it is, the faster the particle's random motions, and the likelier it is to escape the planet's gravitational clutch. Let's consider the examples of water, N_2, and CO_2 (water vapor, molecular atmospheric nitrogen, and carbon dioxide) on today's Earth. For their characteristic speeds to exceed Earth's escape velocity would require temperatures of 90–220 *thousand* Kelvin. Since at about 300 K Earth characteristic surface temperature is vastly lower, very few water, N_2, and CO_2 molecules can attain escape speeds.

This highlights the importance of planets' temperatures, and in general these temperatures drop with rising distance from their stars. But resist the temptation to liken it to the fading light of a lighthouse on a foggy night, which arises from light scattering by liquid water droplets. In the celestial settings, there is little matter between stars and planets to do the attenuation. Instead, planets' cooling with rising orbit radius arises geometrically. Envisioning the star as a point light source from which radiation is emitted uniformly in all direction, for radiation to go beyond r km away from the star in any direction (where r is any positive number) requires doing so in all directions. Therefore, the star's total energy output is evenly distributed over the surface of a whole sphere whose radius is r, on which the planet's orbit traces a great (radius r) circle. And because the surface area of a sphere is proportional to the radius squared, doubling the star–planet distance means fourfold reduction of energy delivery to the planet. Consistently, while Jupiter or Saturn (5–10 times further from the sun than Earth, with corresponding solar warming rates 1–4% of Earth's due to this geometrical dilution) have been too cold and massive to lose even hydrogen, warmer and smaller Earth lost considerable masses of hydrogen and helium.

These losses quickly replaced Earth's primary atmosphere (mostly hydrogen and helium) with a chemically distinct secondary one that has been repeatedly modified over Earth's 4.5-billion-year history, eventually becoming our current atmosphere. Some key modifications, notably the rise in free molecular oxygen (O_2) in the atmosphere, were at least partly life related. Others were life independent, shaped by Earth being relatively cool yet not cold, compact yet massive enough to hold on to most of its molecular oxygen and nitrogen, water vapor, and CO_2, all central to agriculture. The secondary atmosphere has also been shaped not by what wasn't lost, but by what was added. Most important here is volatile enrichment by accretion of meteoritic and cometary material from the outermost solar system[621] and possibly even further away,[7] where low solar heating yields abundant, enduring water, CO_2, methane, ammonia, and sulfur dioxide ices, in sharp contrast to their near absence from Earth's warm inner solar system neighborhood.[9,199,373,555,619,623]

Once delivered, these substances assimilate into the hot mantle, where they volatilize and degas into the atmosphere, modifying its composition. Despite this volatile enrichment, carbon and nitrogen in the bulk Earth (excluding the core, which is essentially fully sequestered and uncoupled from such dynamic Earth affairs as life) are still at least a thousandfold less abundant than the original solar nebula.[242,623] This, of course, is bad news for agriculture.

Further bad news arises from planetary processes that redistribute some key elements within Earth's various reservoirs, thus powerfully modifying material abundances in agriculturally relevant environments, the atmosphere, and the top few meters of the land. Most important is differentiation and core formation, which is intimately related to the Magma Ocean chapter of Earth history,[519] and the roughly contemporaneous Earth's collision with a Mars-sized object that led to the formation of the Moon,[592] degassing Earth and depleting it of volatiles. Magma Ocean refers to an early stage in the formation of such large rocky planets as Earth from small planetesimals (from pebble-size to several kilometers) and planetary embryos (100–1,000 km). Earth's magma ocean period started shortly after initial accretion, and in it much or most of the mantle was molten.[29,169] The heat sources for this widespread melting included radioactivity,[485,521] conversion to heat of kinetic energy of accreting bodies,[519] and—once melting became widespread enough—of gravitational potential energy release by dense blobs sinking through less dense, viscous molten magma.[411]

A key consequence of the above state is the formation of Earth's iron- and nickel-dominated core as metal from impactors sank through the magma ocean (whose density is considerably lower than those of iron and nickel), collecting around Earth's center, eventually reaching a radius of almost 3,500 km (over half of Earth's radius). As this density-based segregation unfolded, sinking liquid metal blobs equilibrated with their liquid silicate environs (the molten mantle) at high pressures and temperatures, yielding the current widely distinct core and mantle compositions.

Most important for our story is that comparing Earth's bulk composition (inferred from its effects on the orbits of planetary neighbors) to seismic records, analysis of meteorites, high temperature and pressure lab experiments, and first principles modeling indicates that Earth's core is some 10% less dense than pure iron–nickel alloy.[30] This implies that the core must contain some light elements. While their presence can greatly impact core physics,[81] from the standpoint of core mass, these are minor "impurities". In sharp contrast, because the core accounts for about a third of all Earth's mass,[625] the sequestration of some of those light elements in the core, even in very low concentrations, wildly reduces their abundance in Earth's crust and atmosphere and thus agricultural availability.

For example, it now appears that not just some, but indeed most (some 90%, albeit with high uncertainty) of Earth's nitrogen is locked up in the metallic core,[289] with about six to eight atmospheres worth of additional nitrogen in the mantle. So, even notwithstanding the uncertain core nitrogen content, no more than roughly a tenth of Earth's nitrogen is present as N_2 atmospheric gas, and this is a tenth of a total nitrogen budget that is already depressed by the nucleosynthetic and accretionary processes described above. And since the main direct source of inorganic nitrogen for life—for all amino acids, peptides, enzymes, nucleic acids, or proteins in all animals, plants, fungi, or microbes—is the atmosphere, the fact that so little of Earth's already anomalously low—relative to both neighboring oxygen or carbon and to the bulk solar nebula—total nitrogen is in the atmosphere as N_2 further diminishes its availability. This adds an important layer to the explanation of agriculture's acute need for nitrogen additions and the dominance of nitrogen over fertilization.

Nitrogen in the Mature Earth

Once cosmochemistry had spoken, determining galactic and solar nebula elemental abundances, and once those abundances then constituted the initial conditions for planetary formation, and once planetary processes redistributed elements within Earth's reservoirs, Earth assumed today's stable composition. Yet "stable" by no means implies stasis, because, like carbon, nitrogen cycles through Earth reservoirs, rapidly in a biological cycle and far more slowly in a geological one.[671]

The biological nitrogen cycle tracks the fast carbon cycle, because photosynthesized biomass includes proteins, peptides, and genetic material, which means that some of it is nitrogen. When that organic matter is consumed by grazers and their predators, this nitrogen is passed on with the rest of the consumed biomass, and some is absorbed and used for the eaters' physiological functioning or assimilated into their body tissues. But all life eventually ends, and sooner or later, all living matter decomposes, and remineralizes. Most prominently, carbohydrates oxidize back to CO_2, amino acids oxidize back to N_2, and both gases return to the atmosphere.

But a small fraction of this biomass takes a major, lengthy detour, the geological part of the carbon and nitrogen cycles. The initial step in the delivery of nitrogen from the fast biological cycle into the slow geological one involves collection of organic detritus on the ocean floor. This unfolds throughout the order 100-million-year journey of oceanic plates, from mid-ocean ridges to to subduction zones. In mid-ocean ridges (nicely exemplified by the mid Atlantic ridge roughly midway between the Americas to the west and Europe and Africa to the east), oceanic plates form from solidifying mantle magma. In subduction zones, such as those along the west coast of South America, they return to the mantle. Over this long journey, the ocean floor collects much organic matter, the remains of earlier now dead oceanic life. Because most bottom seawater is amply oxygenated, those organic remains decompose (rot). But some—especially in oxygen-poor oceanic environments or where rapid sedimentation covers earlier sediments before their organic matter fully decomposes—collects and endures. Over time, these sediments compact, lose interstitial water, and often gradually transform into organic rich rock. When an oceanic plate reaches the subduction zone end of its journey, some of this organic rich rocky material on it subducts back into the solid earth.[417]

Upon subduction, the fate of nitrogen in ocean floor sediments varies greatly by the local mantle conditions, especially the temperature, oxygen levels, and acidity (*p*H) as a function of the rising pressure with depth the subducted material experiences as it enters ever deeper into the solid earth.[398] Subducted plates' temperatures drop with age,[264] which is often a proxy for the width of the ocean basin in question (the greater the distance from an oceanic plate's mid ocean ridge origin, the longer the plate has to cool from its initial flowing magma state in its ridge nursery, see figure 3 of Penniston-Dorland et al.[458]). Subducted plates' temperatures also drop with the speed with which the subducting oceanic plate converges toward the less dense overriding continental plate (the faster this convergence, the less time the oceanic plate has to be warmed by the hot mantle into which it descends, also illustrated in figure 3 of Penniston-Dorland et al.[458]).

Typical temperature differences these dependencies yield, say 400 K, can make a world of difference anywhere along the descent[395,428,671] to whether such volatiles as carbon or nitrogen will be lost by degassing from the subducting plate back into the atmosphere, as their volatilities suggest, or whether the subducting plate will manage to somehow hold on to them instead. For the latter option, the volatiles must be affixed in place as integral components of the subducting rock. The most obvious route to such immobility is incorporation into mineral lattices, the three-dimensional skeletons formed by individual atoms in a mineral form, giving basic mineral units their characteristic shape (e.g., cube, pyramid, prism). Such incorporation can mean substitution—for example, replacing some calcium with magnesium in the mineral calcite (common in limestone, shells, or corals)[186,569] or fitting nicely at a particular spot within the lattice cage as an add-on. For either of those to work, the right fit is obviously required. One aspect of "fit" is size: the replaced object and its replacement must have reasonably close characteristic radii (e.g., the two doubly charged swapped cations mentioned earlier, Mg^{+2} and Ca^{+2}, differ in size by about ±20%). Another requirement for a good "fit" is matching charges and electronic configuration, which guides the magnesium–calcium example, in which both are missing two electrons.

In cooler subduction zones, upper mantle conditions favor ammonium (NH_4^+) as the dominant nitrogen form, and this ion is relatively "compatible" in that it can replace potassium ions in the rigid structures of various mineral lattices, thus minimizing nitrogen loss by degassing. This reflects nitrogen's unique electronic configuration (three unpaired valence electrons, voracious affinity for protons), which dictates that the radius of the five-atom ammonium ion, about 1.5 Angstroms, is actually *smaller* than that of potassium's single-atom ion, about 2.3 Angstroms. By comparison, at nearly 4 Angstroms, the lighter nearest neighbor four-hydrogen molecule, methane, is almost three times larger than ammonium, and thus geometrically incompatible. In sharp contrast, in warmer plates the dominant form of nitrogen is N_2 gas, which is readily lost by degassing into the atmosphere.

However long, the geological part of the nitrogen cycle eventually ends, with a return flux back into the biological cycle. It too takes two key forms. One is mantle degassing by mid-ocean ridge volcanism.[371,372] The other, which we have already encountered, is volatilization and degassing of oceanic sediments on hot oceanic plates subducting in slowly converging subduction zones.[513] In today's Earth, N_2 dominates these fluxes, reflecting the somewhat oxygenated state of the current mantle.[131] If the mantle were oxygen starved instead,[568] as appears to have been the case in the young Earth,[294] nitrogen would have formed mostly ammonium (NH_4^+, in which the oxidation state of nitrogen is -3, reduced). Conversely, if the mantle were better oxygenated, NO_2^- or NO_3^- (nitrite and nitrate, in which N's oxidation states are $+3$ and $+5$, i.e., oxidized) would have dominated degassing nitrogen. Because these are all reactive compounds that bacteria readily convert back and forth, in either case the fate of degassed nitrogen would have been very different, because at least initially (before any further transformations), it would have been readily available for life. In other words, the Earth we now have, with N_2 gas the favored form of mantle nitrogen degassing, is not the only dynamic path Earth could have taken, if it accreted significantly more or less oxygen than it actually has. But taken it it has, the highly inert and biologically unavailable N_2 gas is the favored form of nitrogen in mantle degassing, and life's job became that much harder.

The description in this chapter of the roots of nitrogen's low availability near Earth's surface may have been technically dense at times. Yet its practical message should be clear enough: even the most basic agricultural practices have deep, rich Earth roots. Analyzing environmental dimensions of agriculture without careful consideration of these roots, artificially divorcing agriculture from its cosmic–planetary context, can only deliver incomplete and often misleading conclusions.

8

How Earth's Workings Shape Agriculture

Parts of the food–diet–agriculture–environment nexus are technical, and thus shrouded in considerable mystery for many. This is hardly unique. Just think of our relationship with our digital devices, on which we critically depend, ever more strongly. Most of us without an electrical engineering PhD have basically no idea what goes on behind those shiny screens. Roughly the same also applies to our health; think of those fifty biomarkers whose circulating levels form the rows of the 4 page printout your annual bloodwork yields. Or our home appliances, or our cars, the list goes on.

But, it appears, we respond in widely different ways to various dimensions of our pervasive cluelessness. In some, we simply gracefully accept defeat. For example, with the possible exception of favoring Android or iOS devices, few of us have "opinions" on digital architecture. Most of us similarly humbly accept being entirely at the mercy of our overworked, beleaguered primary care physicians when it comes to interpreting that 4 page printout. But in some dimensions of our ignorance, we fight. Think about the deep emotions stirred up by COVID, its origins, vaccines, Dr. Fauci, masks, and the rest of it. At the height of the pandemic, it seemed like none of us could have held more deeply entrenched opinions on each of these issues, with levels of conviction strongly inversely proportional to how much we actually knew about any of this. The same, it sadly appears, also dominates much of today's food–diet–agriculture–environment discussion. I unfortunately have no idea what explains this curious divergence. But I do know that the option of feigning expertise in the absence of any dominates the popular discourse on agriculture and diet, making most of us easy prey for disinformation and wishful thinking. Despite some heroic collaborative efforts of experts and dedicated skeptical journalists[20,255,407] to rebut naive[429,431,515] or self-serving[290] claims about food, this is where things currently stand.

One major reason for the confusion, I believe, is Earth naivete. Given the multifaceted complexity of the Earth system (some of which we encountered as recently as the preceding chapter), this is completely understandable. To genuinely understand Earth mechanistically, one needs solid graduate-level knowledge of physics, dynamics, thermodynamics, ecology and evolution, botany, biogeochemistry, among other scientific disciplines. Because of this breadth, the ability to synthesize all of it into a coherent picture is also required. How likely are dentists, lawyers, or engineers to possess this kind of disciplinary knowledge, and know how to integrate it into a planetary-level understanding? Yet this system-level geophysical knowledge is precisely what is needed to understand much of what happens in our fields and farms, and the broader environmental dimensions of agriculture.

In light of this, in this chapter, I wish to build on the preceding chapter and provide a *painfully partial* distillation of the essential science that undergirds agricultural human–Earth interactions. Because the scope of this task is impossibly large, I've curated a few key examples that are both of considerable practical societal importance and present key scientific ideas, striving for a tractable, tolerably technical discussion.

The Curious Case of US Dairy Operations Flocking Westward to the Desert

If you are particularly observant about your food, you may have noticed that your milk or yogurt come more and more from such western locales as Colorado east of the Rockies or California east of the Coast Range. There, dairy operations steadily proliferate and assume gargantuan scales, at the expense of such traditional dairy powerhouses as Vermont or New York. To

HOW EARTH'S WORKINGS SHAPE AGRICULTURE

illustrate the point quantitatively, let's compare historical trends in dairy production in New Hampshire and Texas, [439] a traditional New England dairy state and a clear example of the novel dairy frontiers respectively. New Hampshire dairy production has been fluctuating around 180–200% of 2022 production (219 million lb) from the early 1920s through 1962, with steady decline of 1–2% a year since then. Texas dairy production, in sharp contrast, fluctuated around 15–25% of the 2022 value (16.5 billion lb) from the early 1920s through the late 1970s, after which production has mostly been increasing exponentially through the present. The 1923–2022 maxima occurred in 1962 in New Hampshire, but in 2022 in Texas. The New Hampshire and Texas dairy systems have clearly traced diametrically opposite trajectories.

This is peculiar: cattle's evolutionary cradle was cool, lush northwestern Europe. Why would dairy operators force cattle to migrate from lush locales that imperfectly approximate their evolutionary origins to hot, arid places most inconsistent with them?! (Texas dairy is mostly concentrated around its northwest panhandle corner abutting New Mexico, where annual precipitation is under 500 mm, with seventy days a year with temperatures above 32°C or 90°F.)

The reason dairy production used to be distributed, occurring most everywhere, is intuitive enough. Prior to the onset of ubiquitous, affordable refrigeration, dairy production had to more or less track population spatial distribution, so as to meet spatially distributed demand for dairy products, which almost define the American cuisine. Refrigeration liberated us from this constraint. But why did this lead to concentration of most production in a few curiously dry prime regions, with mostly declining production elsewhere? If you noticed this trend, and became sufficiently curious about it, you may have turned to agricultural economics, or the Farm Bill, but likely found the causal explanations they offer inelegant and unsatisfactory.

Here's what actually works: subsidence (atmospheric downward motion, or winds from aloft toward the planet's surface). This may appear curious at first blush; a process involving atmospheric dynamics explaining trends in cattle production? As I show below, subsidence does indeed help explain the mass exodus of dairy production. Appreciating why begins with recognizing the high energetic demands on modern dairy, and the very considerable heat stress dairy cattle are therefore under in the summer heat. [43]

Lactating cows typically generate 25–26 Mcal (105–110 megajoules) per day of metabolic heat, [23] with large, highly productive ones generating more than 44 Mcal (185 megajoules) per day (enough energy to run a typical large window air conditioner for about thirty four hours). For maintaining a steady body temperature, a cow's internal heat generation rate must equal the heat flux through her skin envelope to the surrounding air (aka heat dissipation). Assuming these cow types surface areas are about 5.5 and 6.5 square meters, [51,335] these metabolic heat generation rates require time mean heat dissipation rates of about 220–230 and 330 watts per square meter of skin/coat, respectively. Comparison with human heat dissipation rates helps explain just how large this is. At rest, humans typically dissipate heat to our surrounding at a daily mean (basal metabolic) rate of about 60 watts, or (assuming a mean human surface area of 1.75 square meters) 35 watts per square meter. This rate, which works out to be about 0.86 kcal per kg body weight per hour, [383] defines activity level of 1 MET, resting metabolic rate. Actual metabolic rates of minimally active adults range over roughly 60–75 watts per square meter, rising to 95–110 watts per square meter for adults with physically demanding occupations, and to 170–240 watts per square meter for young adult professional Tour de France bike racers. [460,511] So the metabolic heat burden of a modern early lactation high performance dairy cow is larger than that of even the most supremely trained, uniquely genetically gifted endurance athletes at the highest levels of the most metabolically demanding athletic pursuit!

Several effects jointly cause this outstandingly high heat burden. First, modern dairy cows are highly genetically selected for hyper-performance, yielding 10–12 metric tons of milk annually instead of 3–4 tons per year [164] beef cows do, or the even lower annual yields that must have characterized cattle's evolutionary ancestors. Since producing milk is energetically costly, high performing dairy cows must eat much more than their wild counterparts, and produce correspondingly higher internal metabolic heat. In addition, cattle digestion relies heavily on rumen fermentation, which enhances heat burden per feed calorie beyond monogastric digestion (having one simple low pH stomach, like ours or pigs', instead of cattle's four chambers), with mechanistic details to follow. Third, cattle's large size complicates dissipating internally generated heat through their skin and coat into their environment. This complication arises because metabolic heat generation occurs throughout the animal and is therefore crudely proportional to the animal's mass and thus volume or radius cubed, but dissipated heat must pass through the animal's surface, whose area is proportional to the radius squared. The heat dissipation problem, the ratio of heat generation to surface area, is consequently proportional to the animal's radius; the larger the animal, the (linearly) larger its heat dissipation problem.[227,267,331,384] Owing to their size, cattle are thus at an inherent heat disadvantage. Lastly, much of the developed world uses temperate climate–adapted *Bos Taurus* cattle breeds rather than less productive yet heat stress–adapted tropical *Bos Indicus*.[472]

These processes result in modern dairy cattle being uniquely prone to summer heat stress.[74] One particularly effective remedy for this stress is enhanced sweating, which can powerfully dissipate cows' internally generated metabolic heat to the near surface air environment, *provided* environmental conditions permit rapid enough sweat evaporation. The potential importance of sweating for cattle thermal comfort stems from water's great latent heat of vaporization: the fact that it takes about 2.4–2.45 million joules (or 570–590 kcals, a quarter of a person's daily energy needs) to evaporate a kilogram of water or sweat. Multiply this by characteristic sweating rates of a typical cow, about 0.5–3.5[51,335] kg sweat per hour in typical hot summer conditions (e.g.,[220,567,677] or Eq. 5 of reference 137), and you get 0.3–2.1 Mcal per hour (or, assuming again surface area of 5.5–6.5 square meter, 50 to almost 450 watts per square meter). The upper bound, 2,100 kcals per hour, means that in one hour of effective sweating, a cow dissipates our full daily caloric needs! These high rates are the reason that among all the different cooling processes available to cattle and other large animals, evaporative cooling tends to dominate thermal regulation of heat-stressed cattle.[139,344]

But all of this is perfectly useless if environmental conditions near an animal's surface permit only slow, sluggish evaporation. The dependence of evaporation rates on surface air properties, the key to understanding dairy migration, is illustrated quantitatively by modeled drying time of cattle coats made wet by sweating under various conditions.[110] Drying time at 30°C and 40°C are 16% and 34% shorter than at 25°C (figure 8 of Allen et al.[110]). Drying times at 44% and 29% relative humidity are 26% and 41% shorter than at 58% (figure 10 of Allen et al.[110]). And drying times at wind speeds of 1 and 3 meters per second are 46% and 63% shorter than with no wind.[110] In other words, the hotter, less humid, and windier it is, the faster animals' coats dry. Because temperature, relative humidity, and mean wind often covary, strong winds tend to accompany unusually hot, dry air under some conditions (e.g., the Levantine *Hamsin* or *Sharkieh*), the above drying rate disparities can rise to an order of magnitude or more. This means that the fraction of productive sweating—sweat that evaporates, dissipating heat from the animal to the surrounding air, rather than dripping as liquid—in total cattle sweating can range by tenfold or more. This is important for most animals under heat stress (including endurance athletes in hot environments, hence the custom among Tour de France racers to douse their helmeted heads and backs with bottled water), and doubly important to hyper productive heat stressed modern dairy cattle.

Sweating Favorable Physics

Let's explore the impacts on surface air properties and thus on cattle heat problems of subsidence—downward air motion from aloft toward the surface—in the lower atmosphere (1–5 km or 3,000–16,000 ft above the surface). These impacts unfold in several physically distinct ways. First, we note that water vapor content of air is very low aloft, but rises rapidly toward the surface. In turn, this is because most atmospheric water vapor originates from surface evaporation, and moving stuff vertically in the stratified atmosphere is a lot more sluggish than moving it laterally. Consequently, subsidence introduces drier upper air into low altitude air, drying it. Let's consider the example of climatological (long term mean) seasonal specific humidity (the actual water vapor content of the air in grams water vapor per kilogram of air mixture) averaged over much of the continental US (125°–70°W, 35°–50°N), shown in figure 8.1 as a function of atmospheric pressure, which is high near the surface (bottom of panel) and low aloft (top).

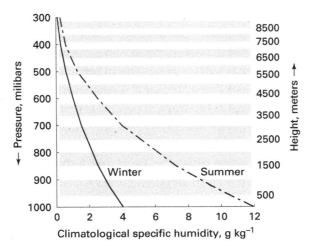

8.1 Observed climatological (long-term mean over decades) air specific humidity between the surface (the plot's bottom) and 9 km above the surface, where atmospheric pressure is 300 milibars (mb), under a third of the surface pressure, where pressure attains its column maximum. Winter (December–February) and summer (June–August) are shown in solid and dash-dotted lines, respectively. In the plot, specific humidity data (in grams of water vapor per kg of air mixture) are averaged over 125°–70°W, 35°–50°N. The gray bands show height range, with the lowermost and uppermost bands spanning 500–1,000 and 8,500–9,000 m above sea level.

The figure shows that aloft, the air contains very little water vapor throughout the year. In both seasons, specific humidity rises downward (toward higher pressure), faster in summer (June–August), and slower in winter (December–February). Clearly, the higher up the air is, the less water vapor it contains, especially in the lower atmosphere, and in summer. Because of this declining moisture with height, subsidence dries the lower atmosphere by introducing into it low water vapor air from aloft. We will quantify this shortly.

Another mechanism by which subsidence 2–5 km up impacts lower atmosphere air properties and thus helps explain the migration of cattle operations to more arid locales involves relative rather than specific humidity. Relative humidity is the ratio of the actual mass of water vapor in air to the *most* water vapor the air can hold, often expressed as percent,

$$\text{relative humidity} = \frac{100 \,(\text{air's actual water vapor content})}{\text{air's maximum possible water vapor content}}.$$

The numerator, the actual water vapor content of the air, reflects the air history: where it came from, how long ago it was in direct contact with the surface, especially of the watery persuasion, and which way the winds that transported ("advected") it blew. The denominator, the maximum possible water vapor content of the air if it had access to such effectively unlimited vapor sources as the tropical oceans, is a basic thermodynamic attribute of air. It depends strongly on the air temperature and weakly also on its pressure. The relative humidity ratio is not an absolute measure of water vapor content (that's the job of specific humidity), but rather a measure of how close a given air parcel comes to water vapor saturation (100% relative humidity), the point at which water vapor begins condensing en masse into the liquid water forms we call, depending on the context, clouds, dew, or fog.

Relative humidity suppression by subsidence arises from the combination of the strong temperature dependence of the denominator in the relative humidity ratio with the tight temperature–pressure coupling in atmospheric vertical motions, as follows. Because atmospheric pressure at a given point reflects the mass of the atmosphere above that point, and the further up in the atmosphere you go, the less mass remains above you (because more is below you), atmospheric pressure drops with height (hence the need for supplemental oxygen in high elevation mountaineering). Consequently, subsiding (descending) air moves from low into ever-higher pressure. In response, descending air parcels that exchange no heat with their environment contract and compress (every kilogram of air occupies less volume) because—in thermodynamic parlance—the environment through which they descend performs work on them, exerting force on them over a certain distance. This work done on subsiding air parcels means that the average kinetic energy associated with the random motion of molecules in them—O_2, N_2, water vapor, CO_2, and so on— rises (that energy and work are both measured in joules highlights their close relation). Since temperature *is* a measure of molecular kinetic energy, descent driven compression means that descending air warms along its downward path. The upshot: air warms as it subsides.

This warming strongly enhances the maximum possible water vapor content of the descending air (its saturation specific humidity, the denominator in the relative humidity quotient given above). This maximal amount is governed by a celebrated fundamental result, the Clausius–Clapeyron relation. It follows straightforwardly from basic thermodynamics of two phases (here liquid and gas) of the same substance (here water) in equilibrium. For the atmosphere, this equilibrium is attained when the mean gas-to-liquid mass flux—condensation of water vapor into liquid water droplets—is equal to the mean flux from the liquid to the gas phase, evaporation of liquid droplets suspended in the air. Given that gas is the more energetic phase (this is why we need heat to evaporate a pot of water, say), it makes intuitive sense that the Clausius–Clapeyron relation dictates that as air warms, increasing its thermal energy, more water molecules can acquire enough energy manifested as speed to vaporize—escape the liquid embrace and strike it on their own as unassociated gaseous water vapor molecules. What is not necessarily intuitive is that this rise is *exponential*, i.e., the water vapor holding capacity of air not only rises with temperature, but rises at a rate that itself rises with temperature. While here and elsewhere I use "hold" to describe the mass of water vapor in an air parcel, it is a misnomer, because at atmospheric (i.e., relatively low) pressures, air is simply a collection of molecules and atoms that coexist passively with minimum, mostly unimportant interactions. So water vapor is not really "held" by the air, but the term is widely used. Returning to and illustrating the exponential nature of the Clausius–Clapeyron relation, the increase in air's maximum water vapor content due to warming from 30°C to 32°C is over triple the increase accompanying a 10°C to 12°C warming (about 3.5 and 1.1 g per kilogram, respectively). This accelerating effect of warming on increasing air's ability to hold water vapor is deeply important to surface relative humidity (again, the dimensionless ratio of the actual water vapor mass the air contains to the maximum mass it can contain at its temperature and pressure) and thus to cattle comfort.

To illustrate both effects in action, imagine, for example, 275K (35°F) air with specific humidity of 3 g water vapor per kg of air undergoing a modest 200 m a day subsidence that induces about 2K a day warming. Initially (at 275K), its saturation specific humidity is about 4.3 g per kg, so its 3 g water vapor per kg specific humidity translates to a relative humidity of 70%. In a single day of subsidence, during which compression-induced warming raises the subsiding air's temperature to 277K, its relative humidity drops to 60% through the effect of increasing the denominator alone.

Let's now combine this with the other mechanism, reducing actual water vapor (the numerator of the relative humidity ratio) content by subsidence. From the dash-dotted specific humidity summer profile, we can estimate that in the lowermost atmosphere, specific humidity drops with height at an approximate rate of 2.8 g per kg per km (which follows from specific humidity of 12 and 5 g per kg at the surface and 2,500 m above it). Multiplying this by the assumed subsidence rate of 0.2 km per day yields a drying rate of about 0.6 g per kg per day. So the specific humidity after one day of this subsidence rate is no longer 3 but instead 2.4 g per kg, yielding a relative humidity of 48%. The relative humidity of the considered subsiding air dropped from 70% to 48% in a single day, a staggering relative humidity suppression rate of a full third per day. This is the power of the two pronged subsidence induced lower atmosphere drying mechanism, combining lowering the actual water vapor content of the air with greatly enhanced capacity to contain water vapor.

We have thus far assembled two of the three ingredients needed to explain dairy migration to arid locales: outstandingly high heat burden of modern high productivity dairy cattle, and subsidence-induced suppression of relative humidity by reducing the air's water vapor content, and by warming it, raising its *maximum* water vapor content. But these subsidence effects act well above the surface; how do they affect the comfort of surface-dwelling cattle and their managers' profits? This highlights two additional ingredients our discussion requires. The main one—the physics that communicate subsidence effects aloft to the surface during summer—will be our next stop. But before that, let's take a brief detour to discuss an essential ingredient in the argument that subsidence affects dairy siting, declining dairy productivity and profits in the absence of a satisfactory solution to the heat dissipation problem. This is an economic and thus much less robust link in our logical chain than the bulk of the discussion, but it is essential. Heat stress easily suppresses cows' milk production by 10–20%, but at times can top 30%,[587] depriving affected dairy farms of a whole third of their normal income. In real well-studied cases, this has sufficed to temporarily turn normally profitable dairy operations into money-losing ones.[562] Add observed recent summers, much longer and more brutally hot than past summers,[276] and concerns for global negative implications for dairy production quickly follow.[435]

Returning to our focus, the physics governing surface drying by subsidence, through physics outlined in detail in the Accompanying PDF, subsidence "caps" the boundary layer (loosely the lowermost atmosphere, where air properties are heavily modified by the surface). This minimizes boundary layer exchanges of energy, water vapor, momentum, and other air attributes with the bulk of the atmosphere above. Appreciating the significance of this begins with recognizing that the atmosphere is *not* warmed primarily by absorbing downward solar radiation traversing it from the top toward the surface, as many intuit, because throughout much of the atmosphere (except some 20–30 km above the surface, where ozone is present at high enough concentrations), air absorbs most solar radiation wavelengths rather poorly. Instead, the atmosphere is primarily heated by the surface, which absorbs solar energy and warms the air immediately above it through thermal conduction. In turn, this warmth is communicated to higher up in the atmosphere via convective vertical mixing, with details in the Accompanying PDF. For thermal energy, the partial isolation by subsidence of the boundary layer from the rest of the atmosphere thus means that the "container" into which surface heat is delivered upward—the shallow boundary layer under subsidence—is smaller than it is with no subsidence, when the boundary layer is much deeper (as comparing panel a in figure 1 of the Accompanying PDF to panels b and c of the same figure reveals).

Let's take a simple example. Suppose a square meter of the surface absorbs enough solar energy to warm the atmosphere above it at a rate of 500 watts per square meter. If this upward heat flux warms a 500 m tall boundary layer, not atypical depth for a boundary layer under strong subsidence, it induces a warming rate of 3.5°C (6.5°F) per hour. If it warms a 2–3 km thick boundary layer, also not atypical for midday summer without subsidence, instead, the warming rate drops to 0.6°–0.9°C (1.1°F–1.6°F) per hour. If this heat flux operates (warms the overlaying atmosphere) between 8 a.m. and 2 p.m., the 2 p.m. air under subsidence is 19–21°C (34–38°F) warmer than in the absence of subsidence (but this is illustrative, not literally predictive, as other processes are sure to limit the warming). Taking a mid-range temperature difference of 20°C (36°F), imagine how dramatically differently we feel at 80°F vs at 116°F (roughly 27° vs 47°C); while the former is an ordinary, potentially even fine summer day, the latter is, even in Tucson, the leading item on the nightly local news. Because after sunrise the summer boundary layer typically deepens, entraining into it air from just above it, subsidence also warms the boundary layer by making that entrained air warmer. Note, parenthetically, that the above can be dramatically mortified, even reversed, in humid environments. But since the new dairy hot spots are all arid or semiarid, we will not dwell on this possibility here.

Subsidence aloft and surface conditions are thus intimately coupled through the physics described above and in the Accompanying PDF, resulting in a warmer boundary layer more isolated from the overlaying free atmosphere. In the absence of a major surface water vapor source—a sizable lake or a large area of heavily irrigated lettuce fields come to mind—the boundary layer under subsidence is also drier. This result appears paradoxical; we set out to explain dairy operations' migration to subsidence-dominated areas, invoking the outstandingly high heat burden modern, high productivity dairy cattle experience. If so, why in the world would dairy operators make a nettlesome problem—their stock's apparent heat stress—*worse* by moving to areas that are made warmer still by subsidence?

The explanation hinges on a simple yet crucial fact: for their thermal regulation, cattle, like us, critically depend on sweat evaporation.[221] In most hot daytime observations, sweating-induced heat loss is some 2–3 times larger than the next largest flux,[678] dominating the thermal exchanges between individual animals and their environment.[159] This is why we focus on conditions favorable to evaporation in the following discussion.

The issue is made clear by figure 8.2. It tracks the evolution over twelve days following the onset of 200 m a day subsidence, which—given the conditions shown in the earlier figure—warms and dries the air by about 2°C and 0.6 g per kg per day. As we have already seen, these changes jointly suppress the air's relative humidity, the warming by increasing the denominator, the maximum water vapor mass the air can hold, and the drying via reducing the numerator, the mass of water vapor the air actually holds. Here, we express these changes as a key determinant of cattle thermal comfort, the "humidity deficit" of the air, or the difference between the maximum mass of water vapor the air can hold—q_s in the vertical

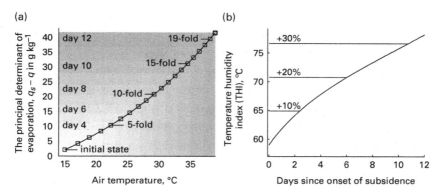

8.2 Panel a: Twelve day evolution of air "water vapor deficit" (see text for details) following the onset of 200 m per day subsidence that warms and dries the air by about 2°C and 0.6 g per kg per day. Time implicitly progresses from day 0, the lower left most square, to day 12, given by the upper rightmost square. Background gray shading helps visualize the passage of time, from the beginning in the lightest gray to the 12th day in the darkest gray. Integer multiples of the initial deficit on day 0 are indicated, culminated by almost 20-fold increases on day 12. Panel b: The evolution over the same 12 day period of an index of cattle thermal comfort, with time presented explicitly along the horizontal axis. The shown THI range can be comparesd to typical daytime summer values in Vermont and Alabama of about 77–79 and 85–86 respectively.

axis label—and the mass of water vapor the air actually holds, q in that same label, both in grams of water vapor per kg of mixed air. While time is not explicitly shown along either of the plot's axes, its passage, from the lower left to the upper right, is represented by the gray shading, from zero (when subsidence first begins; lower left) to day twelve after that (upper right). The squares' spacing highlights integer values of $q_s - q$ relative to the initial state (at time = 0, when $q_s - q \approx 2.1$ g kg^{-1}, given by the lower left white square). Thus, for example, $q_s - q$ becomes ten times the initial $q_s - q$ (marked by "10-fold," about 21 g per kg) some 7.2 days of the onset of subsidence.

The reason the difference $q_s - q$ is of great relevance here is that it is a key determinant of evaporation vigor, with evaporation doubling for every doubling of this difference. To understand why, imagine when it vanishes, when the air is saturated (contains exactly as much water vapor as it *can*, i.e., $q = q_s$, relative humidity = 100%). Then, the water mass flux from liquid to gas (liquid water to water vapor) is exactly equal in magnitude to the opposite flux, condensation of water vapor into liquid droplets. Such equality is the essence (indeed the definition) of equilibrium, and the net transition of liquid to vapor or vice versa vanishes.

From this extreme, the larger the air's water vapor deficit (the further below saturation the air's humidity is, or the larger $q_s - q$ is), the linearly more vigorous evaporation into this air is. And as figure 8.2 shows, twelve days of subsidence raise this difference over twentyfold (>2,000%). That subsidence raises the heat burden (expressed as an index, the THI,[594] that takes note of both temperature and humidity) by no more than 35% (panel b) but at the same time enhances evaporation rate—and thus heat relief by sweating—over twentyfold makes clear why subsidence is such a bargain from the standpoint of cattle heat stress, and why dairy operations flock to locales dominated by subsidence. Put differently, the above, the unique heat stress dairy cattle face in summer, and the singular importance of evaporative cooling by sweating to cattle thermal regulation jointly explain why subsidence matters so much to cattle farmers. To be sure, cattle would more or less always feel more at home in Ireland than in New Mexico. But given that US dairy cannot locate in Ireland, hot and very dry beats slightly less hot and much more humid. This is why the US Gulf States and even New England states steadily lose dairy production.

Before we get into the "why," here's a spoiler: the key is what fraction of the water flux associated with total sweating evaporates vs the complementary fraction that drips off instead, because while the evaporated fraction greatly enhances comfort, the dripping bit only makes matters worse.

Recall that in typical summer temperatures, it takes about 580 kcals to evaporate a kilogram of water (water's latent heat of vaporization). If this heat is derived from the sweating animal, every kilogram of evaporated sweat thus rids the animal of about 580 kcals. If that kilogram all dripped off before evaporating instead, assuming the lost water is at a typical cattle temperature and the animal is steadily replenishing the water by drinking 12°C (54°F) water, the animal loses a mere 26 kcals. Dripping sweat is thus essentially a pure waste, distinguishing the productive—evaporated—from the wasted—dripped—fractions of total sweat production. This partitioning depends strongly on evaporation rates and thus on environmental conditions, as the following end members illustrate. If the air near a cow's skin is hot, its relative humidity low, and a nice breeze blows, evaporation keeps pace with the produced sweat, which all evaporates, productively performing its physiological role of dissipating heat from the animal to the environment. These conditions are the reason it is such a joy to climb up a steep hill in Santa Fe in June, say, despite the high temperature. Even though we are losing water rapidly, our skin (with the possible exception of the back along the spinal column) never gets wet, because sweat evaporates as soon as it is expelled.

The opposite extreme is performing a similar climb in a cloud (called fog near the surface). When I was a faculty candidate at UC Santa Cruz, after my interview ended, I rented a bike in Santa Cruz, rode north a bit on Rte. 1, and climbed—mostly in a cloud—to the Bonny Doon Ecological Reserve. Toward the end of the climb, I emerged from the cloud, and was greeted by pleasantly dry air and a warm, friendly sun that dried me in no time, making me strongly inclined to take the job. Because the air during the climb was saturated with respect to water vapor (its relative humidity was 100%), no net evaporation to speak of took place. In such conditions, unless the air is sufficiently cold to dissipate enough metabolic heat through the skin simply by nonevaporative heat fluxes (there are several physically distinct such processes), hyperthermia quickly follows. But for our current discussion, what happens to our sweat in this case? First, beads form, attached to our skin by surface tension. As long as the beads are small enough, this surface tension overpowers the downward gravitational force Earth exerts on the small liquid mass a bead contains, and the beads persist on the skin. But as the beads grow by continuous expulsion of more liquid sweat through the skin, they become too massive, the gravitational force overpowers surface tension, and the liquid pinches off, dripping off with no actual cooling. The message is clear: such thermodynamically challenged animals as bike riders[37,154] or cattle[152,677] depend critically—sharply more so as heat stress rises (see, e.g., figure 3 of Amano et al.[13])—on effective sweating, which requires rapid evaporation, which requires warm, dry, windy conditions,[111,663] and which powerfully determines their overall thermodynamic trajectory.

The Roots of Cattle's Heat Burden

It is now time to reinsert cattle physiology, with added details, into our Earth physics story, and use the combination in our inquiry into the curious case of recent dairy migration from lush to arid lands. Remember one of the reasons we invoked earlier to explain cattle's anomalously burdensome thermal challenge, digestive reliance on rumen fermentation. At that point, we noted that this enhances the heat burden per unit of usable feed energy relative to what animals that rely on monogastric digestion (having a single low pH stomach, like ours, instead of cattle's four chambers) experience. The key here is the "per unit usable feed energy" qualifier, which becomes all the more important on roughage rich rations.[240,509] A good way to quantify this is the production rate of cellular energy that is made available to the metabolizing animal for performing useful work (e.g., heart beating, walking, digesting) per gram food consumed. And while food (or feed) consumed may comprise any macronutrient—carbohydrates, fat, or protein—the quintessential carbohydrate, glucose, is best quantified and makes the most sense to use.

To quantify useful feed energy, we need to introduce cell energetics and the cellular energy currency, the ATP molecule. Cells derive much of their energy by oxidizing the macronutrients we eat within dedicated structures inside cells called mitochondria. This oxidation, reasonably viewed as "reverse photosynthesis" when the oxidized substrate is sugar, liberates free energy for use by cells and thus the animal, the very solar energy plant leaves used when photosynthesizing the burned sugar. Once liberated in mitochondria, some of this energy must be stored until needed. This storage is achieved by compelling a molecule called ADP to form a high energy bond with an inorganic phosphate molecule, PO_4^{-3}. Symbolically, this energy charging, discharging, and recharging cycle can be summarized as

$$\text{ATP} + \text{water} \rightleftharpoons \text{ADP} + PO_4^{-3} + \text{free energy}$$

(the middle "D" and "T" stand for "di" and "tri," distinguishing the two and three phosphate molecules). Adding phosphate to an ADP thus forms an ATP, and removing it from an ATP forms an ADP, with the "$\rightleftharpoons$" symbolizing this bidirectionality. Availability of free energy (mostly from sugar oxidation in mitochondria) drives the reaction to the left, forming ATP by bonding ADP to another phosphate (for which energy is required, likely for overcoming electrostatic repulsion between neighboring negatively charged phosphates[297]), while consuming free energy drives the reaction to the right. Owing to phosphorus's relative abundance in Earth's

HOW EARTH'S WORKINGS SHAPE AGRICULTURE

continental crust,[597] and the chemical versatility of phosphate,[297,638] ATP is *the* energetic currency of cells and much of life, driving useful biological work from a yeast cell synthesizing a necessary enzyme to an elephant moving her trunk.

Now let's use these facts, and the metaphor of ATP yield per ingested glucose molecule as monetary return on investment, to derive a measure of the metabolic energetic returns (or efficiency) of a given digestion model, energy per unit usable feed energy. The most ATP profitable digestion model is aerobic oxidation, yielding about 30 ATP molecules per glucose (refs. 375, 396 [ID 114969, based on ref. 296]). The least profitable pathway, meanwhile, anaerobic fermentation with no usable daughter products, yields only about 3 ATPs.[277] Of course, reality is somewhere in between, with added ATP production arising from oxidizing fermentation daughter products, primarily acetate, with some propionate and butyrate, which jointly add 11–14 ATPs per fermented glucose (refs. 277, 396 [ID 114969, based on Kaleta et al.[296]), and in reality less, because neighboring microbes can "steal" secreted extracellular daughter products and utilize them for their own metabolism, reducing the fermenters' ATP yield. Another way to view this efficiency disparity is in biomass produced, about 0.5 and 0.15 dry g per g glucose under aerobic and anaerobic conditions. Either way, the fermentation energetic penalty appears to be about two-thirds.

This means that cattle typically derive only one-third to at most one-half of the useful energy from their feed that the aerobic oxidation path would have gotten them. (While this is an approximation, because their feed also comprises fats and proteins that are subject to distinct metabolic and energetic calculus, it is close enough to illustrate the point.) But this great energetic sacrifice is the necessary other side of the very coin that allows cattle to eat grass. That is, while grass is brimming with (repackaged solar) energy, most of this energy is in the form of structural complex carbohydrates, mostly cellulose, lignin, and other polymers of glucose whose digestive breakdown requires enzymes neither cattle nor us are equipped with the genetic blueprint for synthesizing, but that the symbiotic microbes in the cattle rumen are. This energy is therefore mostly inaccessible to the host animal, but available to its microbial symbionts, who use it to drive their own metabolism. Because the energy and protein in the resultant microbial biomass (their little bodies) feed the host cattle and are of central nutritional import to them, ruminants' acceptance of low efficiency turns out to be not a sacrifice at all, but a shrewd bargain, accepting half of the energy with fermentation instead of nothing without it. Cattle could have evolved to either eat grass with the symbioses required for metabolically utilizing it at half the efficiency aerobic metabolism would have provided, or seek other resources (the "nothing" here, what cattle would have gotten from grass had their evolution "insisted" on efficient aerobic oxidation as their sole digestive strategy, jettisoning fermentation). This is approximately true, because most animals host *some* fermentation, but in the hindgut (large intestine, colon), not the rumen. Just ask horses (who are not ruminants, yet who perfectly well subsist on grass because they rely on hindgut fermentation). Even we harbor some fermentation, but while it contributes importantly to our well-being, [50,326,337] its significance to our overall energy budget is small and possibly even negative. [129]

With grazing cattle deriving only 30–50% of the useful energy from a unit of feed aerobic oxidation would have delivered, they must consume 2–4 *times* as much feed as nonruminants would have needed for fueling the same energy consuming activity, such as weight gain or milk production. This explains their enhanced internal heat production, and highlights a built-in constraint on cattle productivity. Since even the hungriest cow can only eat so much grass, some productivity is almost inevitably lost on lower quality grass with low energy concentration. But for cattle thermal regulation, most important is that the elevated roughage consumption the half efficiency requires enhances internal heat production per available calorie. This would have spelled trouble for any animal, but it is particularly burdensome for cattle, who are already saddled with a disproportionately large thermal challenge, as we have seen.

Before continuing, let's recap the detail-rich yet fundamentally simple message of this chapter thus far. First, regulating body temperature and avoiding overheating are uniquely challenging for cattle, all the more so for high productivity dairy cattle and forage eaters (and cattle *must* eat at least some forage). Second, sweating is potentially particularly effective at ridding cattle of excess heat, *provided* surface conditions permit rapid evaporation, without which sweating becomes useless.

Westerlies, Mountains, Subsidence

As we've seen, by warming and drying the lower atmosphere, subsidence accelerates near surface evaporation particularly effectively. This is very useful for understanding cattle's thermal challenge, but it raises a key question: What partitions Earth's surface into either subsidence-dominated settings in which evaporation is very fast, or locales with little or no subsidence, in which evaporation is often too slow for sweating to realize its full cooling potential? To answer it, we apply the full temperature–pressure–vertical motion–evaporation rate package we just developed to the dynamic interactions of large-scale wind systems with mountains. Because this discussion is motivated by dairy operations migrating to the westernmost Great Plains, the prevailing winds we will focus on are the westerlies (west-to-east winds) that characterize the midlatitudes (and thus most of the contiguous US), and the mountains we will focus on are the Rockies.

Once the westerlies encounter the western slopes of the Rockies, they are diverted from their original path vertically and—because the Rockies' peaks range widely in elevation along the range's north-south expanse—laterally. These vertical and lateral displacements trigger restoring forces. For vertical displacements, the restoring force arises from buoyancy differences: air that was forced upward expands and cools faster (about 10K per km) than the environmental air, thus finding itself amidst warmer air it is denser than, sinking back down, reversing its initial forced lifting. For lateral displacements, the restoring force arises from the fact that the eastward speed of Earth's surface due to Earth's spinning about its axis of rotation varies smoothly between an equatorial maximum and zero at the two poles. The nature of this restoring force is less intuitive, but is not essential for the current discussion. What *is* essential is that the vertical and lateral displacements that arise from the interactions between the midlatitude westerlies and the Rockies induce a unified restoring force. It propels air toward its original pre-disturbance position, downward for upward displacements, or southward for northward displacements. Because of inertia, the resultant motions overshoot, with upward displaced air parcels descending below their original height, or northward displaced parcels going further south than their original latitude. Which is to say, such generalized displacements excite waves, not unlike the way plucking a guitar string away from its resting (minimal tension) position compels it to generate sound by oscillating through the air.

The most coherent impact of the excited atmospheric waves is strong subsidence just downstream of the range. For the westerlies' interactions with the Rockies, "downstream" means "east of", and subsidence is maximized over the westernmost Great Plains. (If you want to really get it, I warmly recommend the relevant discussion (around pp. 542–545) in Geoff Vallis's wonderful *Atmospheric and Oceanic Fluid Dynamics*,[613] where figure 13.1 beautifully shows the stationary wave response described qualitatively above.) A good measure of the coherent downstream manifestation of this wave is the observed long-term annual mean subsidence some 3 km above sea level (at 700 mb) near the Colorado–Kansas state line (38°N 103°W), about 290 m per day. Let's now quantify the two main effects of this subsidence over the western Great Plain. Subsidence drying of the lower atmosphere by introducing into it air with low specific humidity (less water vapor per kg air) is quantified by multiplying the subsidence rate by the rate at which specific humidity declines with height. Some 3 km above the westernmost Great Plains, this product yields a mean dehydration of about 1.8 g per kg per week. To get a sense for just how vigorous this drying rate is, note that in under two weeks, dehydration by subsidence stands to completely eliminate the entire resident specific humidity present at that location, 3.5 g per kg. While other processes will eventually curtail this rapid dehydration, it is enormous.

If this comparison wasn't clear to you, here's a hopefully clarifying analogy. Imagine your lettuce patch is infested with caterpillars. You have two clear options. You know pesticide application works, but it would contaminate your lettuce with toxic residues. Introducing a natural caterpillar eater to the patch is also possible, but its efficacy is less sure. How do you quantify the two outcomes and make an informed choice? One way is to compare the number of caterpillars the introduced caterpillar eater can be reasonably expected to eat per day to the current number. Suppose the current infestation is one hundred caterpillars per square meter, and the introduced caterpillar eaters can eat two caterpillars a day; all the lettuce will be long gone before a meaningful number of caterpillars are eaten. On the other hand, if you can reasonably expect the caterpillar

HOW EARTH'S WORKINGS SHAPE AGRICULTURE

eaters to take out 200 caterpillars a day, well now you perk up, because if you introduce them after breakfast, by lunchtime your patch will be caterpillar free. This comparison is analogous to comparing the dehydration rate to the long term mean specific humidity, with the rate of caterpillar consumption standing for subsidence induced dehydration rate, and the current one hundred caterpillars per square meter standing for the long term mean specific humidity. In general, if the flux into or out of a reservoir of interest is fast enough to change appreciably the reservoir within a time period of interest, then the flux is important to the reservoir. That is the relationship we found above for the dehydration rate and the local resident specific humidity.

This suppression of the air's specific humidity or water vapor content—the numerator in the relative humidity ratio—by subsidence is further amplified by the second effect, reducing relative humidity by increasing the denominator in the relative humidity ratio, the air's *maximum* water vapor holding capacity, by subsidence induced compressional warming. The 290 m per day annual mean subsidence reported above warms the air by 2.8K per day. At the annual mean temperature some 3 km above the westernmost Great Plains, 276 K, this 2.8 K per day of subsidence warming enhances saturation specific humidity by 1 g per kg (from 4.6 to 5.6 g per kg) per day. Given the annual mean specific humidity there reported above, 3.5 g per kg, this 1 g per kg per day saturation specific humidity enhancement suppresses relative humidity by over fourteen percentage points daily. In about six days, the cumulative impact of this relative humidity suppression exceeds the long term mean relative humidity there (84% vs. 78%). Combining subsidence's warming of 2.8 K and drying of 0.26 g per kg per day, and applying them to the local air 3 km above the westernmost Great Plains, with annual mean temperature and specific humidity of 276 K and 3.5 g water vapor per kg, results in a staggering local relative humidity drop of 18% a day! Based on the relative humidity dependence of cattle coat drying time discussed earlier—1.1–1.2% faster drying per 1% drop in relative humidity—the 18% a day relative humidity reduction means that drying time drops by 80% in a week, with one minute of drying time on Monday becoming twelve seconds by the following Sunday night.

With this, we are done with the process-based explanation of why, seemingly inexplicably, dairy operations flock to the least probable destinations, the westernmost Great Plains, the Central Valley of California east (downstream) of the Coast Range, and similarly arid locales that intuitively seem most inhospitable to moist- and cool-loving cattle. To combat cattle's exceptional heat burden due to their large size, high performance, and large metabolic heat output, migrating to where sweat dries fast, thus dripping minimally, is essential. This requires near-surface air characteristics that strongly favor rapid evaporation. Because few processes promote these conditions more than subsidence, which outside of the tropics is maximized downstream of mountain ranges that vigorous prevailing winds pass over, modern dairy migrate to those areas.

With all of these processes and their concerted action understood, things fall into place in an intellectual catharsis that is an earthly analog of Dobzhansky's[155] famous statement that "[n]othing in biology makes sense except in the light of evolution," with agriculture standing for biology and Earth physics and cattle physiology for evolution. Far from perplexing, locating dairy operations just east of the Rockies or the Coast Range now makes perfect sense, because that is the location of maximum effect of the planetary waves the interactions of the prevailing westerly winds with the mountains excite. And in that most coherent downstream node, the main effect for our purposes is subsidence; more subsidence, more comfortable and productive dairy cows. One imperfect and indirect way to show this involves comparing dairy productivity in two classes of states: traditional humid New England dairy states (Connecticut, Massachusetts, New Hampshire, Vermont), and more arid, emerging dairy powerhouse states (California, Colorado, Kansas, Texas). Observed 2015–2022 mean milk productivities in those state groups were 9,500 and 11,000 kg milk per cow per year.

With respective variabilities within the two groups of 640 and 520 kg per cow per year, the arid states' 16% milk yield edge is statistically significant, with fewer than two shots of a hundred to arise randomly. More importantly, this is economically revelatory. The reason is that by far the main income stream for dairy operations is milk sales (accounting for 90–93% of all their income in the US in 2022), and the main non-labor expense is capital costs of machinery and equipment (accounting

HOW EARTH'S WORKINGS SHAPE AGRICULTURE   141

for 83–87% of all nonlabor costs).[226] Notice the dichotomous difference between these two leading items; whereas the principal income source is directly linearly proportional to milk production, the leading nonlabor expense is independent of milk production, depending instead mainly on operation size, the number of cows milked. To see why the latter dependence holds, note that one of the most capital intensive elements of modern dairy operations, typically second only to buildings, is the milking parlor. Parlor size, and thus the investment size it represents, is strongly proportional to the number of cows milked. If a dairy operation's main income stream is proportional to the milk mass sold, while the size of the possible investment strongly discourages expanding head count, clearly the most obvious option is to increase milk yield per head. For that, one better devise an effective strategy for handling summer heat burden. One such strategy is relocation to where sweating is very efficient. And of all approaches to enhancing sweating efficiency, few match the ferocious efficacy of subsidence. This is indeed what US dairy producers do, in earnest, explaining a great deal of the impetus to migrate to the western Great Plains. While not all of the above 16% milk yield edge of arid states is attributable to aridity, much of it is, and much of the remainder arises from economies of scale, which conditions subsidence establishes also strongly promote.

On Apples, Eastern Washington, and Subsidence

Let's now build on the physics we just learned to explore another curious agricultural phenomenon: distribution of US apple production. While Washington and New York dominate 2007–2022 US apple production statistics,[439] they trace very different trajectories. New York's share ranged over 8–14% of the total US production, essentially unchanged from its 13–15% share in 1949 and 1959 (recorded by the respective editions of the U.S. Census of Agriculture). Over the same period, in contrast, Washington, almost exclusively east of the Cascades, accounted for a whopping 54–72% of the US total apple production, a rapid jump from its 18–23% share in 1949–1959. Before development of advanced refrigeration, longhaul food transportation, and ripening on demand via controlled atmosphere storage,[283] apple production (like dairy production we analyzed earlier) was distributed, occurring most everywhere, lest regional shortages would arise. But once these technologies matured, liberating the food system of this constraint, productivity became a key factor, with simple market forces strongly favoring geographical specificity that optimizes growing conditions. As in the preceding dairy story, over time, this led to concentrating production in locales that are nearly ideal for a given food, with declining or vanishing production elsewhere, even in places where conditions can be reasonably described as "almost ideal."

But why is Washington, especially the Yakima Valley, just about apple nirvana? Here's a clue: the June–December long term mean relative humidity in Yakima, Washington, never exceeds 50%, and cloud cover is 10–15%; daytime is mostly warm or hot, sunny, and pleasantly dry. Compare this to the Hudson Valley, a New York apple powerhouse, where climatological mean cloud cover during the same months never drops below 35%, the relative humidity is in the 75–100% range, and it rains about half the time. In part, this is because summer total atmospheric column water vapor masses in eastern Washington and southern upstate new York—observed by NASA's MERRA-2 satellite[223]—are very different, about 5–15 vs 35–40 kg per square meter, respectively. (The units envision hypothetical precipitation of all water vapor in an atmospheric column of interest as rain, with each kilogram per square meter covering the surface with a liquid water layer one millimeter thick.)

Hot, sunny Washington summer days are ideal for minimizing pests, blights, and disease that can exact a heavy toll on apples in such humid areas as New York's Hudson Valley. Summer nights also differ greatly in Yakima and the Hudson Valley, mostly due to already mentioned low cloudiness and atmospheric water vapor content overhead. At a basic level, the reason is simple enough. Water vapor dominates the atmosphere's greenhouse capacity,[523] the surface warming due to intercepting by atmospheric constituents of upward infrared radiation from the surface, and the enhanced downward infrared radiation that follows and that warms the surface. Consequently, low water vapor content over eastern Washington lowers overhead greenhouse capacity and hastens nighttime infrared radiative cooling. The surface impacts of clouds are more nuanced, depending on such variables as cloud heights (mostly as a temperature proxy, with high clouds being much colder), droplet size distribution, among others. But at night—when incoming solar radiation vanishes and thus shielding the surface from some of it, as clouds do during the day, is immaterial—clouds, like atmospheric water vapor, mostly further warm the surface. The low cloud cover over eastern Washington thus further promotes nighttime cooling, as does the fact that most apple orchards around Yakima are 200–500 m above sea level. Since, as we have already seen, atmospheric water vapor content drops rapidly with height, atmospheric water vapor mass is disproportionately concentrated near sea level. An elevated local surface therefore excludes from the local atmospheric column the layer that would have otherwise contained most of its water vapor, resulting in a smaller total overhead mass, and more nighttime cooling.

Cool nights, relatively short and mild winters, and early spring onset all make eastern Washington ideal for apples in another way, generous cumulative annual chill-hours—the number of hours with temperatures between freezing (0°C) and 7°C—that many apple varieties, especially the coveted Honey Crisp, require. In the Hudson Valley, the longer, harsher winters, and the month later spring onset limit winter accumulation of chill hours by keeping temperatures too low, while rapid spring rise of daily low temperature—propelled by low topography and high water vapor burden—does the same in spring by promoting earlier onset of nighttime temperatures above 7°C.

What makes the Yakima valley so ideal for apples is thus its sunny, warm, and dry summer days, and its cool nights, and the underlying cause for those is, once more, being located just downwind of the Cascades. As before, the interactions of the prevailing westerlies with the Cascades excite the same waves as those we encountered earlier, with similar results: strong (and at times stronger, see, e.g., the left panel of figure 5 in Overland[447]) subsidence, strong air dehydration and compressional warming, depressing relative humidity, and re-evaporating any stray cloud that may have been brave enough to try to form.

To estimate the magnitude of these processes, we start with observed summer subsidence rates averaged over July-Septmeber, 1995–2005, 121°–120°W, and 46°–47°N. Estimates naturally vary, but one[491] that explicitly resolves the diurnal cycle and represents well the key effect here, topography and the upstream presence of Cascades, reports mean subsidence of about 0.4–0.7 km per day some 3 km above the ground. This yields representative rates of warming by compression due to the increased pressure into which the subsiding air descends of about 4–7K (7–13°F) a day. The second hydrometeorological effect of this subsidence, dehydration, is given by the observed specific humidity vertical gradient there—about 1.4 g per kg per km on average[298]—times the above subsidence rate, about 0.6–1 g per kg per day.

Next, let's quantify the effect of these two drying mechanisms on the local air, whose mean summer specific humidity and temperature are 2.8 g per kg and 277 K (39°F). Given that saturation humidity at 277K is 5 g water vapor per kg air, 2.8 g per kg corresponds to 56% relative humidity. Combined, 4–7K per day warming from 277 K and dehydration of 0.5–1 g per kg per day from 2.8 g per kg suppress relative humidity by 21–35 percentage points per day, with relative humidity dropping from 56% to 21%–35% in a day. These rapid drying rates propagate, with modification, to the surface (remember the boundary layer vertical mixing we discussed), positively affecting apple health and reducing pest infestation. [58,362,668]

Such positive effects have clearly not been lost on apple producers, giving rise to eastern Washington's dominance over US apple production, even over the second largest player, New York. This is more general than apples alone, because any agricultural enterprise—any crop or livestock—has its ideal physical conditions, and thus geographical locations.

We can speculatively imagine the process that culminated in Washington's apple primacy, and—more broadly—in the sharply honed geographical specificity of modern agriculture. It could have started, before the proliferation of long haul transportation, with spatially distributed production that closely tracked demand (population). Because agronomic suitability varies widely geographically, distributed production necessarily meant a broad range of yields, quality, and needed inputs (such as labor, agrochemicals, capital investment, or transportation). These disparities likely propelled corresponding regional disparities in consumer food prices, farmers' cash receipts, and farm income. The dawn of affordable long-haul transportation, food transportation in particular, around and following World War I[636] must have played a key role in changing this state of affairs.

It is not unreasonable to expect progressively lower food transportation costs—which in recent decades accounted for a mere 3–4% of US real food prices[91,379]—to promote spatial homogenization of diets and their costs.[600,622] This likely spelled shrinking profits for orchards in suboptimal geographies, where yields are lower and the need for agricultural inputs potentially higher. This rising geographical variability in profit margins likely propelled the geographical specificity we have analyzed in both dairy and apple production, with the unsurprising "winner takes all"—or at least more and more—outcome we have identified in both.

While this scenario is clearly speculative, it offers two testable predictions: large disparities in revenue streams should now exist between ideally and less ideally situated farms producing the same product, and production should be progressively more concentrated in the hands of fewer, more dominant players. Is apple information for New York and Washington consistent with the first prediction? Multiplying the mean 2021 apple yields, 34 and 45 metric tons per hectare per year, by the corresponding prices received, $600 and $675 per metric ton,[439] yields cash receipts of $20,400 and $30,400 per hectare a year in New York and Washington. So yes, Washington enjoys a whopping 50% financial edge over its nearest competitor, New York, with much of it reflecting Washington's inherently better suitability for apple growing.

Good test cases for the second prediction—increasing concentration of production in fewer, more dominant hands—are widely popular foods whose demand can be assumed to closely mirror population, so that if their production proves highly geographically localized, this cannot be attributed to similarly localized demand. For the American diet, oats and turkey are two such foods. In 1920 and 1940, 48 US states produced oats, but this number dropped to 44, 36, 29, and 24 in 1960, 1980, 2000, and 2020.[439] There is no turkey data for 1920, but in 1940, 1960, 1980, 2000, and 2020, the number of states with appreciable turkey production was 48, 48, 32, 25, and 14.[439] Another food that fits the bill is sweet cherry, which 9, 8, 3, and 3 states grew nontrivial amounts of in 1998, 2008, 2018 and 2022 (the USDA sweet cherry records only go back to 1998).[439] So yes, production appears to steadily concentrate in fewer, more dominant hands.

Another imperfect yet suggestive relevant observation addresses trends in land allocation to wheat in two groups of states. The first group—Kansas, North Dakota, Texas, Oklahoma, Washington, and Colorado—comprises relatively sparsely populated dry wheat powerhouse states, while members of the second—New Jersey, New York, Missouri, Ohio, Pennsylvania, and Wisconsin—are lush, humid populous states. Relative to the 1922 value, the wheat area of the first—wheat ideal—group was 6–7% and about 11% higher in the 1940s–1980s and 2000s–2022. Conversely, the wheat area of the second group was 4%-6% and about 7% *lower* in the 1940s–1980s and 2000s–2022.

The trends presented above are clearly consistent with our predictions. While they do not "prove" my conjecture, neither do they disprove it. And when a conjecture is confronted with data, the most the data can do is disprove it, which they did not do here. Moreover, while undoubtedly speculative, the proposed scenario makes sense. It just might be true.

What to Do about Agricultural Regionalization

The vignettes in this chapter tell a straightforward story. Combining Earth's geographical variability with highly specific, narrow physical conditions optimal for different agricultural organisms (crops or livestock types) results in widely ranging regional suitability for these organisms. While most locales can harbor most crops and livestock types, for each, few geographies are ideal, while most are adequate yet suboptimal to various degrees. Because modern agricultural profit margins are very low, regional optimality—which entails lower input needs and higher yields—confers considerable relative advantage. It is logical to then assume that this two-pronged profit-enhancing mechanism is how we got to the high regional specificity that characterizes modern agricultural production today.

This raises a practical question about the environmental wisdom of "buying local." Proponents emphasize reducing the energy use and greenhouse gas and pollution emissions associated with food transportation, and the economic and environmental benefits (other than emissions; more on those in a minute) of distributed production models. Opponents, meanwhile, highlight production phase optimality—low required inputs, high yields—the above regional specificity offers, and the small contributions of transportation to total adverse environmental impacts of food (e.g., food transportation accounts for only 3–5% of total food emissions[133,418]), concluding that buying local is a missed opportunity to exploit the geographical "perfection." From this, an opposite practical recommendation follows: buy each product from the geography for which it is best suited, thus further rewarding naturally gifted geographies and enhancing geographical specialization. Which of these opposing ideas should guide our behavior?

You know the answer you get from a scientist. It depends. Mostly, this goes back to a recurring theme in this book and in environmental discourse writ large, the metric dependence of the relative merit of every choice. Here's what this means. Recall the many physically distinct adverse environmental impacts of agriculture we discussed earlier. Recall also that while some choices improve matters across the board, the most obvious example being reducing one's needs for all resources by replacing beef, most improve matters in some ways while worsening them in other ways. The choice between the above two options is emphatically of the latter kind; both choices have pluses and minuses. Let's briefly explore this.

Favoring local food often emphasizes lower "food miles," how far food travels, sometimes weighted by the emissions characteristic of each transportation type (such as rail, barge, or small truck). Without the latter weighting, "food miles" is almost meaningless, because different distances and different cargo types call for different transportation modalities that differ greatly in their emissions per ton-mile (the transported food's mass times the distance traveled). The best known anomaly here is air freight, that at 500–2,500 g CO_{2eq} per 1,000 kg-km, towers over all other transportation modalities as the uncontested worst transportation emitter.[134,324] Because this is also reflected in higher costs, air freight accounts for a minuscule fraction (under 0.2%) of total food transportation.[163,475]

The implication is as clear—avoid flown food—as it is impractical, because consumers can rarely identify air freighted foods. If your supermarket (truthfully) labels its foods' origins, it is a relatively safe bet that asparagus, green beans, or berries with origins beyond 80–100 miles away have been flown in, and are best avoided. More generally, very short shelf life foods are best gotten locally regardless of where you fall on the above dichotomy.

Beyond air freight, emissions are a poor rationale for favoring local food, a classic example of simple arithmetic trumping lofty ideals for guiding dietary choices. One reason is, once more, the need to focus limited goodwill and budgets on reducing consumption of the largest current resource users, because this stands to yield the most reductions per unit effort. In addition, the mode of transportation can matter far more than distance traveled. For example, long haul freight trucking emits on average about thirteen times the emissions by rail[441] and eighteen times the emissions by barge, while local light truck delivery emits about fivefold more than heavy long hauling.[165,204] The only emission-related benefit of favoring local food is therefore avoiding flown food (which is, remember, a tiny fraction). For food hauled by any other means, production-related emissions are so much larger than transportation emissions that production efficiency is the key consideration. Since that efficiency is greatest in the ideal geographies covered above, consuming foods grown in their respective ideal locales is often the choice that minimizes emissions.

But there *is* one persuasive rationale for favoring local food. Let me introduce it with a little anecdote. My wife loves soft shell crab, and used to order it on special occasions. Nowadays, special occasion or not, you can hardly find them. Where have all the crabs gone? Succumbed to anoxia, one of the best observed triumphs of our tenacious war against Chesapeake Bay. And what is our key weapon in this war? Fertilizer-rich agricultural discharge, mostly from livestock operations,[400] as described earlier in this book. Let's not belabor yet again the virtues of avoiding animal-based foods, and simply consider a person who has already settled on chicken for dinner. For her, the only remaining question pertinent to the current discussion is whether to favor a local chicken or the most efficient one, which, as we have seen, is in general nonlocal. Assuming she lives anywhere outside of the Chesapeake Bay watershed, if she chooses local, she redirects some nutrient-rich effluent from the Chesapeake water to wherever she, and her local chicken, live, reducing her contribution to Chesapeake Bay anoxia and to my wife's want of soft shell crabs.

The general moral of this anecdote is that if you primarily wish to reduce your contributions to disrupting biogeochemical cycles, and are less concerned with your emissions, then local may well be the choice for you. If widely replicated, a distributed local food consuming population would result in farmers using fertilizer and discharging its unused effluent more geographically uniformly, thus reducing nutrient burdens on such currently heavily nutrient polluted bodies of water as the Chesapeake Bay. If, conversely, reducing emissions is your primary concern, which seems logical given the enormity of the climate woes we are in, your best bet is consuming every food from the locations it is best suited for. Since this is the modern supermarket default, this option in effect means "don't sweat it".

9

From Sunshine to Food

Eating is a social, nutritional, and environmental act. For me, subjectively, most central are the environmental dimensions of eating. They weave us most intimately into our planetary, cosmic neighborhood, and that is my emotional home. While perhaps a tad uncommon, and likely unpopular, this view is the primary prism through which I examine eating and, more broadly, nature. This is not only because it feels comfortable, but just as importantly because it delivers; solid, reproducible explanations of our and agriculture's relationship with the natural world. Not only are these explanations practically useful, they are *elegant*, opening windows into fundamental planetary processes that are—like in the dairy and apple examples of the previous chapter—rarely explicitly recognized in the food and agriculture discourse, and even more rarely satisfactorily explained mechanistically.

To be sure, the planetary perspective on agriculture is not necessary; agricultural economists, barley breeders, or mindful chefs can all achieve their unique goals based on taking it as a given, with few whys and hows, that sunshine availability won't limit spinach productivity in the Central Valley of California, that excessive rains won't clobber a southeastern Idaho wheat field, or similarly well observed, deceptively simple but actually fundamental Earth characteristics that affect agriculture. But by forgoing unexamined givens, and revealing the underlying physics, the planetary perspective on food is far more illuminating, satisfying, and beautiful, following paths that are not only exciting but also likely to yield fewer errors.

In keeping with this *Earth First!* perspective, in this chapter we deepen our exploration of the planetary perspective on food using a particularly prominent example, corn for livestock feed, the poster crop of industrial agriculture. About 36 million hectares, almost a quarter of all US cropland, are allocated to such corn, vastly more than all cities, towns, and hamlets put together. As any casual observer of the American road will tell it, corn dominates the American rural landscape, and is the reigning monarch of US agricultural exports. Corn is thus a worthy object of inquiry, and this chapter is about understanding the basic Earth roots of corn's dominance, and using this knowledge to illustrate broader fundamental dilemmas about agriculture and its environmental impacts.

Let's start by briefly pointing out a few key but likely mostly unrecognized ways in which basic Earth processes powerfully shape global corn production.

Since corn obviously needs water (lots of it, as we will explore shortly), Earth's size and composition (the bulk abundance of various elements in Earth's crust and atmosphere) matter. Together, they determine, and are determined by, Earth's mass. This dictates, as we have seen, its gravitational field, and thus a planet's ability to retain light volatile elements and water over the billions of years of its history. A hypothetical Earth much nearer to the sun, or a much smaller one even at the current distance from the sun, would have likely retained less water than the one we have.

But the gravitational field of the Earth we have not only facilitated water retention long ago but also powerfully controls current precipitation, both total amounts and distribution in time and space, through its iron grip on the large-scale circulation of Earth's oceans and atmosphere, atmospheric stability, and the basic characteristics of atmospheric waves (notably the spatial extent and duration between successive individual high and low pressure systems we call weather, and with which most midlatitude precipitation is associated). All of these determine winds, precipitation, temperature, and water vapor (humidity) characteristics,

FROM SUNSHINE TO FOOD   153

and the geographical patterns of their long term aggregates, climate. Imagine a Midwest with twenty to fifty days between rain events, say, instead of today's two to five days; that's an alternative-Midwest free of rainfed corn, and it is possible on a starkly differently sized Earth. You can perfectly well say, therefore, that corn is acutely aware of Earth's gravitational field, even if its farmers may not be.

Gravity is not the only hand at the weather helm, dominating the geographical distribution of corn. Another is Earth's rotation rate, which—if markedly changed from its current value, a twenty-four hour rotation—would also completely alter our notion of weather. What could have changed Earth's rotation rate? Today's rate bears the signature of Earth's heavy iron-dominated core, which is a frozen testimony to a time in Earth's infancy when much of the planet was molten rock, allowing heavy iron to sink to the center and lighter minerals to float and dominate the crust, where we and corn dwell. If Earth did not undergo this "differentiation," its spin rate would have been much slower, because more of its mass would be located further from the axis of rotation than is true today, and days would have been considerably longer. This would have altered the rate at which perturbations—like the ones that arise from the prevailing westerlies encountering the major mountain ranges we encountered earlier in the dairy and apple vignettes—are swept eastward by Earth's midlatitude westerlies (their Doppler-shifted propagation). This too would have impacted Earth's corn hospitability, because this propagation rate determines how frequently a weather system passes over a given point on Earth's surface on average, which also controls the characteristic pacing of rains, and depends sensitively on Earth's radius and spin rate about its axis.

Highlighting different timescales and different physics is corn's obvious need for land, blanketed by thick soil. That current land masses are disproportionately in the northern hemisphere explains this hemisphere's dominance over today's corn production. But it's worth keeping in mind that this is not a permanent state but an ephemeral random snapshot of the state of Earth's restless, ever-shifting plate tectonics. Corn's need for thick soil highlights another important planetary process, variations of Earth orbital characteristics. One such variable characteristic is how close to a circle the annual ellipse Earth traces as it orbits the sun is. Eccentricity is the relevant measure, with 0 indicating a perfect circle, and values closer to 1 indicate more strongly elongated elliptic orbits. Today, Earth eccentricity is about 0.02, with smallest and largest Earth–sun distances of about 147 and 152 million km, but it has varied over Earth's history from almost a perfect circle (eccentricity = 0) to as high as 0.06, triple today's.[627] Also critically important is the axial tilt, the angle between a line perpendicular to the surface of this ellipse and the Earth's spin axis (roughly connecting the two poles) from 90 degrees. This angle, about 23° today, determines how different midlatitude summers are from winters, and varies over roughly 22°–25°; if it were 0° (if Earth's spin axis were exactly perpendicular to the plane Earth's orbit around the sun traces), there would have been no seasonality, and when it is higher than currently, seasonal differences are more pronounced. This angle too varies over tens to hundreds of thousands of years.

Here is why corn cares about these planetary parameters.

The focus here is on the mineral (as opposed to biologically derived) components of soil, is essentially crushed weathered rock. One of the best ways to rapidly produce massive amounts of crushed rock begins with covering a large land mass, such as the central US or the European interior, with thick ice sheets typical of glacial periods. When such ice sheets disintegrate as another glacial period draws to a close, massive armadas of mile-thick glacial ice traverse the land down topographic gradients en route to the nearest ocean. You can well imagine the colossal rock grinding and thus mineral soil formation that occur at the bottom of these oceanward flows; this is the mineral basis for such major modern agricultural provinces as the US Midwest or Ukraine's vast wheat, canola, and sunflower areas.

To get massive amounts of fine ground-up minerals we need for soil formation, we thus need ice ages to ebb and flow. To get glacial periods, you need land on which to place vast masses of ice (sea ice is very thin and too fragile for the job). Tropical lands are, for the most part, irrelevant, because they bake under too much incoming solar radiation, so that even if—improbably—it snows in winter, the snow fully melts come summer. Instead, you need mid- to high-latitude land masses on which winter snow cover may persist through seasons, years, and eons, but relatively cool summers are a must. There are different ways to get cool summers. Here's one.

FROM SUNSHINE TO FOOD

Imagine a time in Earth history when Earth's orbit around the sun is more elongated (eccentricity further away from zero) and the axial tilt is lower than today, say 22°. The latter will suppress mid- to high-latitude seasonality, cooling summers there somewhat by reducing solar radiation (the incoming rays are more oblique, lower above the southern horizon in northern high latitudes) where ice is most likely to first accumulate. This minimizes summer snow melt, retaining more ice. In other words, we are poised to enter an ice age, *provided* the random distribution of land masses prepared a nice nursery for our future ice sheet.

The elongated ellipse can help, or interfere, with our plans, depending on where Earth is in this elongated orbit during summer solstice (currently June 21st–22nd in the northern hemisphere) in the land-dominated hemisphere (which is currently the northern one, but which also varies over Earth history). Again, because snow melt risks our nascent ice sheet most in summer, if summer in a land-covered hemisphere occurs when Earth is furthest from the Sun (aphelion), it further cools summers. Today's state is rather close to this, with aphelion trailing northern summer solstice by as little as two weeks.

Some combination of the above ideally suitable conditions prevailed during the last glacial period, but this ended some twenty thousand years ago or so, bringing an abrupt end to the last glacial period. The mineral components of the soil we enjoy today in such key agricultural provinces as the US Midwest or some of the Great Plains are the products of the demise of the last ice sheet, ground by receding ice. Clearly, understanding corn and its ideally suited production areas requires appreciating planetary processes. Let's now delve deeper into corn's ubiquity and a few of the fundamental processes on which it depends.

At first, corn's hegemony seems almost obvious: our most basic task for crops is to grow, and corn outgrows most other crops. Like all plants, but more uniformly and predictably, and *much* faster, corn plants take up CO_2 gas from the atmosphere into their green leaves, and use solar energy to photosynthetically convert it to carbohydrates and other plant matter (biomass) we use as food or feed. The following numbers illustrate this. US corn yields about 40–50 million kilocalories (or kcal, of which a typical adult needs about 2,000–2,500 daily) per hectare annually when processed into either kernels or silage (which uses the full aboveground biomass of younger, still green corn plants). By comparison, yields of potatoes, rice, apples, barley, soybeans, almonds, wheat, oats, spinach, and oranges are roughly 35, 31, 22, 14, 14, 12, 10, 9, 3, and 1 million kcal per hectare per year. A hectare of corn thus yields anywhere from 10–40% more to 28–36 *times* more calories than other crops on average, a prodigious biomass producer matched by very few crops. But there is much more to diet than calories, and biological productivity requires much more than just sunshine. These needs are our next stop.

A road map for the remainder would help. We began by comparing corn productivity to that of other crops, revealing that in terms of dietary energy yield per acre, corn is almost second to none. The next stop addresses the high water and nutrient costs of this high productivity, noting that water and nutrient supply lines must not only be large enough (but not too large) over seasons and years, but also steady, with no wild fluctuations. We then link the adequacy and regularity of these supplies to soil characteristics and climatic regimes, whose joint geographical patterns powerfully explain corn's. We conclude by highlighting how the overlapping geographies of specific climate patterns, soil attributes, and super-productive corn directly reflect basic planetary parameters and processes.

What Corn—and Food—Need

Because CO_2 is the photosynthetic assembly line's main input, flux of atmospheric CO_2 into leaves, the seat of this assembly line, must be steady. This gaseous CO_2 uptake is made possible by mouth-like pores—stomata—that connect the leaf interior with the environmental air around it, which most leaves have many millions of in their outer layers. Each stoma is equipped with a pair of lip-like guard cells whose internal liquid water pressure controls, Cerberus-like, the opening or closing of the aperture. When guard cell internal water pressure is high, the stoma they guard is open, permitting free gas exchange between the internal leaf environment and its surrounding air. Conversely, when the internal pressure drops, the guard cells deflate, and the stoma closes. Because when photosynthesis is underway CO_2 is constantly used up to form carbohydrates inside the leaf environment, partial pressure of CO_2 inside the leaf is lower

FROM SUNSHINE TO FOOD

than in the atmosphere. (Partial pressure of a gas in such low total pressure gas mixtures as the atmosphere is the pressure it would have exerted if its particles were effectively isolated, interacting meaningfully with no other gas particles. Systems in which total pressure is very nearly the simple sum of partial pressures of all constituents, such as the atmosphere, are so-called ideal gas systems.) This CO_2 partial pressure difference drives a passive diffusive CO_2 flux into the leaf (where "passive" means "occurring spontaneously" or "requiring no energy investment"). This steady invasion is the net carbon flux required to sustain photosynthesis.

So far, it seems like the plant is hands down the winner of a perfect marriage of physics and biology. But there is more. Because healthy leaves contain liquid water–filled vascular networks, water vapor often exhibits the opposite behavior, with partial pressure inside leaves typically much higher than in the air around them. This yields a down gradient diffusive water vapor flux, called transpiration, in the opposite direction to carbon's—that is, a loss from the leaf to the air around it. While CO_2 passively invades leaves, water vapor thus steadily hemorrhages out stomata and into the surrounding air. Because stomata allow carbon into leaves, they permit photosynthesis. But in a classical Faustian evolutionary deal, they also permit water losses that leave plants at the mercy of sufficient soil water availability, without which steady spontaneous uptake of liquid water through the roots ceases. The opposite carbon and water fluxes between leaves and their surrounding surface air are thus inextricably coupled.

In lush environments such as tropical rainforests or the Gulf of Mexico coast in summer, water loss through stomata affects plants minimally. First, the high humidity of the surrounding air means that water vapor partial pressure inside leaves is only minimally higher than in the surrounding air, minimizing water vapor loss. In addition, there is plenty of rain to replenish whatever little water vapor the plants do lose. But in dry environments, like southern Israel or the westernmost Great Plains, this stomatal duality imposes a real constraint on plant productivity. There, drier surrounding air means markedly enhanced water vapor losses, which meager rains replenish sluggishly. When the plant consistently loses more water through transpiration than its root system can deliver from the soil, it wilts, guard cells deflate, and stomata close. This stems the unchecked water losses, but its inevitable price, cessation of atmosphere-to-leaf CO_2 flux and thus of photosynthesis, is swiftly exacted. This is why allowing corn to do its productivity magic requires steady resupply of water, in most corn regions around 500–700 mm per season (5–7 million liter or 1.3–1.8 million gallon of water per hectare per season) but as much as 4–5 times that amount in such arid places where corn is curiously grown as Israel or Arizona.

Biological productivity also requires uptake of nutrients, inorganic forms of elements life requires, with nitrogen—as, for example, nitrate (NO_3^-)—often heading the list in agricultural settings. Typically taken up by terrestrial plants from the soil as solutes dissolved in water, nutrients are among the principal direct or indirect links connecting life with the inorganic world: rock, water, atmosphere (notably N_2, O_2, or CO_2).

Why are nutrients necessary? Plant growth requires new cells. All cells comprise proteins and nucleic acids, molecules made of smaller subunits whose biological functionality is derived from strategically located nitrogen groups. Ergo, building living matter requires adequate supplies of biologically available nitrogen. The "biological availability" qualifier alludes to an apparent paradox: whereas all plants bathe in nitrogen (as the dominant atmospheric gas by mass, N_2), suggesting limitless supplies, nitrogen availability strongly constrains plant productivity. This manufactured paradox disappears once you recognize the very stable triple bond connecting the two nitrogen atoms in atmospheric dinitrogen. For example, comparing the energy required to break the triple bonds in N_2 and P_2, the former is almost double the latter. Because of the exceptional stability of atmospheric N_2, it is essentially inert, behaving almost like a noble gas that stubbornly resists forming other chemical associations that would render it biologically available. If plants had feelings, I imagine, this would have been as maddening to them as thirst is to castaways in a lifeboat bobbing on a tropical ocean.

To become biologically available, the triple bond at the heart of this behavior must therefore first somehow break. Some of this breaking occurs by brute-force physics in such hyper-energetic environments as air surrounding lightning bolts, or even in exhaust pipes of car engines. But most naturally occurring nitrogen fixation—the processes of rendering N_2 biologically available—is propelled by gentle physiological temperature

FROM SUNSHINE TO FOOD

biochemistry, mediated by a handful of microorganisms that have the genetic code for building nitrogenase molecules. These enzymes catalyze the disassociation of N_2 gas, followed by association of each of the two liberated individual nitrogen atoms with three hydrogen atoms apiece, thus becoming biologically available ammonia, a special case of the broader class of "reactive" nitrogen molecules, ones that can be the nitrogen source for plant life.

The principal pathway by which growing plants acquire nutrients in general and reactive nitrogen in particular (i.e., participate in the biomass production race) is as dissolved substances hitching a ride on the general soil-to-leaf water flux through plant roots and vasculature (the liquid transporting tissues in vascular plants; think of those roughly circular structures you see in the cross-section of a freshly cut celery stalk). Therefore while plant needs for water and nutrients are quite distinct in mechanisms and biological origins, water and nutrient supply lines are merged. No amount of soil nutrients can satiate plant needs and sustain productivity without adequate water to mobilize them for root uptake, and no amount of nutrient-free pure water can sustain plant growth.

Water and reactive nitrogen are by no means the only substances corn hyperproduction requires, because many elements—headlined by phosphorus and potassium, but also including trace amounts of many heavier elements—are also required. All must combine adequacy with stability. Adequacy addresses seasonal to multi-year timescales, over which total supplies must replenish plant usage that is not returned locally as decomposing litter. Stability—never deviating too much for too long from the mean supplies adequacy requires—is required because most plants can only store relatively small portions of their daily or weekly water and nutrient needs, so supplies must regularly replenish exhausted stores over these timescales. For example, if your begonia loses through leaf transpiration v liters of water per summer, you must provide it with v liters every summer. But you cannot drown it in v liters at once, but spread it over, say, 20 watering sessions over the summer, in each dispensing roughly $v/20$ liters, modified as needed to account for recent temperatures, humidity, or cloudiness.

For field plants like corn, the sole source of water, reactive nitrogen, or any other of their needs is the soil in which they grow. And where does the soil get its supplies of these vital substances? The heavens above, the rock underneath, and higher topographies that drain toward the location in question. Most soil water originates from precipitation. Similarly, most soil reactive nitrogen is atmospheric N_2 gas, microbially split and combined with oxygen, hydrogen or carbon into reactive nitrogenous compounds dissolved in water. Beginning with their journeys through the landscape, up root systems and stems toward the leaves, the plant nutrient and water supply lines therefore merge. This combines the flow within and among Earth's various reservoirs (atmosphere, soil, groundwater) of water with that of life-essential, mostly light elements (like carbon, nitrogen, or phosphorus). From the atmosphere, through the surface, soil, underlaying rock, and down topographic slopes they jointly travel, sharing common pathways and thus fate, as they doggedly traverse the continents, ever oceanward.

Resource Variability and Crop Productivity

The partly overlapping journeys of life-sustaining water and nutrients to plant root systems are thus coupled, circuitous, eventful, and easily disrupted. They are parts of a complex web that connects Earth's oceans and atmosphere, the land surface, the shallow solid earth underneath it, water and nutrient fluxes through and among these media, and terrestrial primary productivity, agricultural in particular. This interconnected web is key to terrestrial productivity, and thus to food production. It is well worth briefly describing here.

A central variable in this web is the landscape's ability to retain water. To get a sense for this, imagine two patches of land—one covered with lush, waist-high prairie grass, the other a newly built subdivision in which roofs, roads, and driveways dominate—twenty-four hours following a slow and steady rain event. In the intact prairie, the grass cover would use up much available water for a productivity burst of rapid growth, and regulate the flux of water to the surface by intercepting falling rain, thus enhancing slow and steady water percolation into the soil while minimizing runoff. With all its impervious surface coverage, meanwhile, the developed plot would yield substantial runoff.[160,231] The dichotomy in this thought experiment illuminates a real issue of great import to

FROM SUNSHINE TO FOOD 161

agriculture. Because the intensity, duration, and frequency of rain events vary widely, meeting the requirement for stable water availability introduced earlier depends crucially on the ability of the surface and subsurface to hold on to rain water. If too much of the precipitation runs off, too little remains behind and becomes available for resident plants until the next rain event. An extreme example of this is the fate of rainwater following a downpour in New York City in August. Within an hour nearly all water is already in the artificial drainage system, on its way to the Hudson or the East River, the few remaining puddles are quickly drying up, and by the following morning, the sad urban trees already need Park Department watering.

This reality seems to bode well for traditional agriculture that occupies lands with deep soils whose voluminous pore spaces can store large volume of excess water percolating down from the surface. But this can easily turn into too much of a good thing; if root systems stay inundated for too long, even by a day or two, roots begin to rot. The response to this illustrates a fascinating general trend of declining tolerance for deviations from the "ideal" norms in modern agriculture, where everything is carefully titrated to near productivity-maximizing perfection. This applies to water or nutrient availability, as well as to weed or pest infestations, and casts an ever-larger portion of all naturally occurring events as intolerable aberrations that must be—as is so common in agriculture—technologically fixed. And if we agree that they are broken, agricultural engineering's ability to technologically fix things is awe-inspiring. If you ever took the time to *really* examine a modern combine, or a baler, or a cotton picker, you know, deeply and viscerally, that while John Deere or Case stocks may not be as sexy as those of Microsoft or Facebook, the human ingenuity achievements of agricultural engineering are lightyears ahead of anything those latter bastions of technological mediocrity will ever achieve. But as is typical with all technology, agricultural engineering also sometimes assumes a decidedly mad aspect. Which brings us to the modern practice of tiling. Tiling, as in mining or dining, mind you, not tilling.

In the US, much of the Midwest is tiled. In the mid-twentieth century, that meant literally being underlain, some 1–2 meters below the surface, by large tiles, a magnified version of your bathroom tiles. Today, those have largely been usurped by perforated pipes, a thinner rendition of our leach fields, for those familiar with rural home sewer systems. But whereas in those systems wastewater flows from within the pipes toward the soil environment, the flow in tiling pipes is reversed, from the soil environment into the pipes, quickly draining fields and directing their exess water downslope, toward irrigation ditches, creeks, rivers, and ultimately the ocean. But, you may wonder, doesn't the proverbial farmer pray for rain? If the prayers are answered, isn't undoing them by speeding up rainwater flow from fields to rivers to ocean therefore shooting ourselves in the foot? Not necessarily, if you consider root system rot. This practice completes the hijacking of the hydrological cycle by modern agriculture. When it is too dry, for more than the tolerated duration (oh, some twenty minutes, generously), irrigate. And when it rains, drain. Fast. That's what tiling strives to achieve, demonstrating the length to which modern agriculture will go to dampen deviations from desirable norms.

As exemplified by the prairie–subdevelopment comparison, natural or imposed land cover changes can alter significantly the near surface hydrological–biogeochemical multiconnected web. Natural examples include wildfires or the emergence of destructive novel pests, while anthropogenic ones include climate change, plowing, tiling, urbanization, de- or reforestation, or agricultural land conversion such as repurposing cropland for extensive grazing. All can markedly modify physical surface and subsurface characteristics such as soil organic carbon content, bulk density, percolation rate, or vegetation coverage, altering transpirational water loss, and direct evaporation from the soil. All of these changes ultimately modify the partitioning of locally available water into the retained fraction vs runoff. Because if water isn't retained, so are the dissolved nutrients it contains, the modifications all also impact nutrient balance on scales ranging from local to catchment or basin scales that may span—as in the case of the Mississippi basin—continental scales.

Turning from local storage to local consumption, plant water needs depend on the vigor of photosynthesis, which also changes. One way it does is in response to varying available photosynthetically active solar radiation, the part of the incoming solar radiation that plants energetically exploit to propel photosynthesis. Some such variability arises naturally, notably by the eleven-year sunspot cycle, which reflects internal solar physics and produces small, predictable changes in photosynthetically

FROM SUNSHINE TO FOOD 163

active radiation. Human action can also perturb photosynthetically active solar radiation in sometimes unexpected ways.[629] One example arose in response to declining emissions of various ground-level pollutants due to stricter air pollution standards. The resulting observed reduction in atmospheric aerosol load (or, with slight oversimplification, haze) caused marked declines in diffused solar radiation (sunlight that hit atmospheric particles on its way down and was diverted from its original direction, thus arriving more or less equally from all directions), reducing biomass photosynthetic production and thus water and nutrient uptake by plants.[308]

Cleaner air, which is of course societally enormously beneficial and saves many lives annually, can have other effects as well. In an unforeseen twist, it has been depriving the lower atmosphere of its earlier abundant supply of cloud condensation nuclei, small and hyperactive particles (such as imperfectly combusted soot) that accelerate the condensation of atmospheric water vapor into liquid water droplets in clouds. Because low-level clouds shade the ground, their recent reduction has yielded more downward solar radiation flux. To be clear, this unexpected outcome is *not* a strike against air pollution standards, which are an unquestionable life saver, but a simple example of a process that can modify the flux of solar radiation incident on the surface and thus the biosphere's water uptake rates.

Better known and of potentially greater societal concern are agricultural impacts of changes to more familiar climate variables like temperature, surface winds, and humidity, arising from either intrinsic natural climate variability or human actions. The most intuitive agricultural manifestation of climate change is the establishment of new climatological norms for temperature, precipitation, and related atmospheric variables that collectively subject crops to conditions novel enough to establish new yields that differ markedly and coherently from historical ones. For instance, if a corn field that had been historically annually yielding 200 bushels per acre on average starts receiving consistently less rain, all else being equal (same farmers, same farming techniques), it will likely deliver less mass and fewer calories and protein.[114,679] In addition to changing long term means (addressing the adequacy criterion), climate change is also about new statistics—expected magnitude, frequency, or likelihood— of up or down deviations from long term means. In fact, the most readily observed agricultural impacts of climate change are related to failure to satisfy crop needs for stability, not adequacy. But both types of change impact both the supply of water to the landscape and the rate of its removal from it. These altered rates jointly feed back into the lower atmosphere and the larger climate system, sometimes dampening and in other instances amplifying the original changes. [8,194]

Because environmental conditions—both long term mean conditions and those during distinct weather events and such "freak" events as hurricane landfall, tornado, or unusually powerful storms—strongly impact crops, climate impacts the current large-scale structure of global agriculture. Climate change—including changes in both expected norms and the brief deviations from them—is therefore likely to modify this structure in the near future, [210] updating the geographical distributions of key agricultural provinces.

Soil Conditions and Farming Decisions

Environmental perturbations of any duration beyond a few hours modify crops' productivity in mechanistically multifaceted ways. Some, such as those of surface air temperature or humidity, affect crops through the air surrounding them. Others, notably such direct physical air-to-ground assaults as hail storms, or deep freeze immediately following rain that are often associated with fast eastward sweeping deep midlatitude cold fronts, also undermine crops' aboveground parts physically. But many are also communicated to crops by the soil in which they grow. This highlights the prominence of soils to agriculture and partly explains the widely appreciated observation that like climate, soils—the last stop in the atmosphere-to-crop journeys of water and dissolved nutrients, from which crop roots draw these essential resources directly—powerfully control crop performance. In addition to soil being the plants' home, much of this decisive role stems from the wide ranges of composition, physical properties, microbiomes, and biogeochemical functioning soils exhibit. Depths of soils or their mineral precursors (unconsolidated crystalline mineral rock fragments) range from a few hesitant millimeters to hundreds of meters. Soil organic carbon content—which strongly impacts the nature and rates of soil-water interactions and thus water and nutrient delivery to resident crops—spans virtually zero to hundreds of grams C per

FROM SUNSHINE TO FOOD

kg soil. Soil bulk densities range from as little as 200 kg per cubic meter to nearly that of continental rocks, about 2,800 kg per cubic meter. Importantly, much of this range reflects variability in pore space per unit soil volume, space that can temporarily store water and solutes (nutrients in particular) and that promotes microbial processing of these nutrients into substances suited for uptake by crop roots. Soil physical properties thus vary over one to at least six orders of magnitude (ten- to at least a million-fold, the latter being roughly the width ratio of the contiguous US and your closet). Because of this range, soils can match superbly the needs of a particular crop, be utterly incompatible with them, or, most often, fall in between.

A key element of the crop–soil compatibility puzzle is that soils do not merely transmit to resident plants news on weather and climate fluctuations, pests, or human insults to the crops, but interpret them. Like Op-Ed columnists interpreting already known events rather than reporting new ones, soils filter the aboveground messages by subjecting them to their physical and biogeochemical properties. Nowhere is this more relevant than in the case of the concentration, composition, and chemical state of nitrogen-bearing compounds in the pore water underlaying a fertilized field, which all steadily and markedly change with time after fertilizer application. Much of this change is mediated by soil microbiota—mostly bacteria, fungi, and viruses—which steadily modify these nitrogen-bearing molecules. Most important to our current discussion is denitrification, in which some soil bacteria modify (reduce) nitrate (NO_3^-) ions, typically using electrons liberated during respiration (oxidizing and decomposing organic matter back to CO_2). While a fundamental natural process, this has major environmental implications in agricultural settings, because nitrate reduction "neutralizes" biologically available nitrogen by converting it back into biologically inert N_2 gas.

On the face of it, rendering reactive nitrogen inert by denitrification sounds like a terrible bargain for crops and farmers: we fertilize fields, at great environmental and financial costs, to provide resident crops with reactive nitrogen they need for growth, but by undoing this, denitrification robs crops of that nitrogen. What's left out of this picture, however, is that much of the applied fertilized mass, typically about half,[352,666] is not taken up by roots, becoming instead the principal water polluter.[531] If a reactive nitrogen molecule is in field soil, it can still be taken up by crops and enhance productivity. Once downstream of the field, between the nearest irrigation ditch and the ocean, however, this molecule can no longer enhance agricultural productivity, but is very likely to cause pollution by nutrient overburden. Denitrification effects on the used and unused portions of the applied reactive nitrogen are thus opposite. Denitrifying unused effluent nitrate exacts no cost—neutralized molecules have no potential benefit—but likely reduces water pollution. Denitrifying field nitrate, on the other hand, competes with productive use—promoting plant growth—of environmentally and financially costly reactive nitrogen, and, worse yet, does so *after* production costs—fossil fuel use and emissions associated with industrial fertilizer production and application—have already been paid.

The coupling of surface hydrology and biogeochemistry introduced earlier now combines with these opposite effects to thrust hydrological changes due to land modifications—faster drainage due to tiling or urbanization, say—back into the story. The key here are rates. The first is the range of typical water flow rates through the field in such units as cubic meters per day or gallons per hour, say. This range is greatly increased by tiling and mechanical tilling (by, e.g., a plow, disk, cultivator, rototiller), and by favoring annual over perennial crops. Another important rate characterizes reactive nitrogen uptake by the resident crop of a given field (the nitrate mass per hectare their roots take up and deliver to the aboveground biomass in a day, say). A third important rate is the denitrification rate characterizing the resident microbiome (kilograms of denitrified nitrate per hectare per day, say). These rates jointly powerfully dominate fertilizer use efficiency, the fraction of total applied reactive nitrogen that is productively utilized by being incorporated into crop tissue. In turn, this also determines agricultural emissions and water pollution. The faster water vacates the field, and the slower applied reactive nitrogen (minus losses to volatilization or microbiological denitrification) is taken up by the field's resident crop, the more applied reactive nitrogen discharges unproductively. And of the unproductively utilized reactive nitrogen, the slower downstream microbiota denitrify discharged nitrogen, the worse downstream pollution gets. Similar principles also govern the fate of such contaminants as potentially pathogenic bacteria or viruses

FROM SUNSHINE TO FOOD

we introduce into the soil via irrigation with imperfectly treated reclaimed water or liquid manure application.

Soils thus modify—attenuate or amplify, extend or shorten—local response to externally imposed environmental fluctuations (e.g., droughts, deluges, heat waves). Well functioning soils can neutralize many perturbations, allowing crops to recover with relatively modest penalties. Degraded or compromised soils, or those that are ill-suited for their allotted use, however, can turn such events, or even smaller ones, into emphatic season enders or even longer and more painful environmental, agricultural, and human disasters. This is especially true when unusually large and often coupled deviations occur from respective norms of temperature, precipitation, humidity, cloud cover, ground-level ozone, precipitation pH, or airborne dust content, among many other possible environmental perturbations. In exceptional cases, such as the Dust Bowl event that terrorized the southern Great Plains in the 1930s, the primacy of soil is overwhelming. The human and natural-physical dimensions of the Dust Bowl are discussed in countless publications. Two personal non-technical Dust Bowl favorites of mine are Timothy Egan's *The Worst of Hard Times*[166] and Ken Burns's *The Dust Bowl* 2012 documentary. Let me give you a brief synopsis. The essence of the Dust Bowl—whose natural and societal dimensions were all predicted and vociferously warned about[585]—was a combination of an unusual but by no means unprecedented naturally occurring drought, made dramatically worse by a series of mutually compounding human errors. These mistakes were mostly related to agricultural practices, especially a heedless rush to join a temporary wheat bonanza due to elevated demand during World Word I that all but dried up once the war was over. To join this frenzy, homesteaders needed to deep plow their land, which was pervasive enough in the 1930s Great Plains to merit a dedicated term, "sod busting." One of the key impacts of this plowing was to aerate the prairie soils, promoting rapid oxidation of soil organic matter. This severely undermined the complex biological–physical–chemical web that the root system of the perennial prairie grass is, and thus the hydrological–biogeochemical functioning of vast swaths of the Great Plains, swaths covered in among the finest soils anywhere. No longer cohesively held together by organic matter and thriving soil biogeochemistry and microbiota, the desiccated unconsolidated mineral particles left behind were easily lofted by the notorious plains winds, eventually all but destroying not only the pre-Bowl local agricultural economy but indeed much of the pre-1930 human community of the Great Plains.

Beyond modulating the severity of impacts of imposed perturbations on crops, soils can also stretch or truncate these impacts in both time and space. To illustrate the impact on duration, let's consider the fate of water a summer storm rained on a field. Relatively sandy soils permit the precipitated water to rapidly percolate to deeper soil layers even if the storm is heavy and persistent. If soil depth, porosity, and structure jointly also offer ample void space, this percolated water can collect below ground (with actual storage also modified by precipitation rates, prior saturation, or topographic slope). By minimizing water loss to downslope flows, this also reduces topsoil loss by erosive surface runoff. Taken together, these impacts allow the soil to extend the productivity enhancing phase of the precipitation event while minimizing its potential adverse effects. Roughly the same is also true, for somewhat different reasons, for organic rich heavy soils with thriving microbiota that constantly work the soil matrix, maintaining complex connectivity of ample pore space. Particularly important in this case are the chemical and surface interactions of organic particles with water, also enhancing water retention and extending its availability in time beyond the rain event itself.

Desert soils rich in fine silt or clay, on the other hand, may form a nearly impenetrable hard pan at or near the surface almost as soon as rain begins, redirecting the lion's share of rainfall into rapidly lost unproductive runoff or shallow, rapid seepage. In this case, most water runs off (sometimes leading to deserts' notorious and notoriously dangerous flash floods), and is therefore only available briefly in the uppermost soil during the storm itself,[224,454] and minimally in later times of need and to deeper roots or microbiota. Such drainage therefore reduces crops' productive use of precipitated water and overall water availability, and promotes erosive loss of essential topsoil, minimizing the benefits of the precipitation event while maximizing its potential adverse effects.

This dichotomy combines with climate to powerfully impact the global distribution of agricultural production. For example, in the US, the

Midwest–plains boundary that separates dominion of corn and soy to its east from wheat, other small grains, and grazing to its west (traversing north-south Kansas, Nebraska, and the Dakotas) closely tracks not only precipitation—more eastward, less westward—but also the boundary between Udoll soil to the east and aridity consistent Ustoll to the west, and, further west, between tallgrass to the east and shortgrass to the west. In space too, soils can transmit local events further afield.

A useful, tragic historical example of this process in action is again offered by the Dust Bowl event. During that dreaded decade of drought and suffocating dust storms, some farms were intermittently buried under a thick coat of deadening sediment that was lofted from rouge upwind neighbors' abused lands. When a storm persisted for too long or coincided with a critical time in a crop's life cycle, it meant the end of another futile season. This process also self-perpetuated because once the crop on a given field was ruined, the field was abandoned, likely then becoming a source of more lofted dust.

To recap, crop productivity in general and corn uber-productivity in particular require: a suitable climate, with weather characteristics— notably temperature, humidity, wind, and precipitation fluctuations over hours to weeks—that are within the location-specific expected norms. Together, these must translate into steady and adequate but not excessive water and nutrient availability, ample sunshine for photosynthesis, and low enough water stress to permit a sufficient number of open stomata, photosynthetically active hours in the day. These climatic conditions must overlap with having the right soil for the job. This means having the right size distributions of mineral particles and organic components, and microbiota compatible with the climate, weather, and the needs of the chosen crop. It also means sufficient soil depth, porosity, and pore size distribution to store sufficient water and nutrient within reach of the root systems of the desired crop. The storage capacity per unit area the soil depth, porosity and composition jointly yield must then match the precipitation regime and the crop. For example, if the typical spacing of precipitation events is five days, the soil must store about seven, and preferably ten or more days' worth of the water needs of the resident crop at any developmental stage and under all reasonably expected weather events, thus allowing it to safely coast over most reasonably expected precipitation lulls. Finally, excess water that may periodically arise in the soil must be adequately drained over no more than a few hours to a few days by an established network of creeks and minor tributaries with stable banks. In most landscapes, this in turn requires some but not excessive topographic gradients, and thriving vegetation along streams. Next, we need to understand what makes Earth so uniquely able to meet these needs.

10

Smooth Earth Operations: Reservoirs, Fluxes, Cycles

If you ask Earth, agriculture is about rearrangements: concentrating in production units (a field, a hen) more than is naturally occurring of some things, while diluting others, all for enhancing useful removal. Here is what this means.

Thinking of "agriculture" stream-of-consciousness style, what images spring to mind? For most, "irrigation," "fertilization," or "harvest" are safe bets, and amateur gardeners would surely add the bane of our existence, "weeding." While clearly distinct, these actions share an important commonality; all modify natural fluxes into or out of the functional unit rather than introducing new ones that are absent from its natural state. Take the irrigation example, choosing a field as our (nonunique) functional unit. Rather than a total novelty, irrigation adds to naturally occurring fluxes. Into the field, those include precipitation, and runoff and soil seepage from higher ground. Naturally occurring outfluxes from the field include evapotranspiration (recall, the water loss that combines direct evaporation from the soil with water vapor loss through leaf stomata associated with photosynthesis), and runoff and seepage to lower ground. Or, take a dairy cow as our functional unit. Grain- and legume-rich rations greatly augment the energy and protein influxes she would have enjoyed in unmanaged environments, while yielded milk and manure, the primary outfluxes, are greatly enhanced by generations of careful selection. Manure spreading similarly augments naturally occurring nutrient fluxes into a field and the energy, protein, and mass fluxes in harvested biomass out of it. All of these fluxes would have existed even if the field were a natural untamed grassland, forest, or wetland, or if the cow were a wild auroch. But they are all modified, typically greatly enhanced, by agriculture, with the deliberate, explicit objective of enhancing the flux on which farmers are singularly focused, harvest. (If you are straining to reconcile harvest with a flux, think about the yield units, bushels of corn, or tons of canola, per hectare per year, say, or kilograms of milk per dairy cow per year. That these are all canonical flux units makes clear that harvest is indeed a flux.)

The above examples highlight several fundamental recurring elements of agriculture. One is conserved attributes—energy, linear and angular momentum—or substances—water, nitrogen, carbon—whose comings and goings are readily quantifiable. Another is a set of distinct, interconnected pools or reservoirs such as a field, an agricultural province, and all Earth components—atmosphere, biosphere, ocean—they are connected with. And the third element is the bidirectional exchanges among those reservoirs, fluxes of the conserved attributes or substances (e.g., ocean uptake of atmospheric CO_2). Because masses of Earth's individual elements, with the minor exception of hydrogen, are fixed, nothing is lost during material cycling. These three components thus define cycles—of carbon, nitrogen, or water, say—whose existence distinguishes static moons or spent planets from living ones. (The meaning of "spent" will become clear shortly.) This structure, which agriculture shares with most natural biogeochemical and physical processes on and in Earth, further highlights the many inextricable ways agriculture is embedded in the broader but mostly private lives of planets.

To keep such cycles going invariably requires energy sources. Even apparently spontaneous fluxes often aren't on closer examination. Take, for example, the seaward flow of water precipitated over land, which seems to simply follow topographic gradients and thus release rather than consume energy. But if precipitation were to stop, these oceanward flows would quickly die down as well. Precipitation over land must therefore endure, which requires water vapor to reliably converge onto the atmosphere overlaying land. In turn, this requires winds to carry water vapor around, and winds bump against mountains, or rub against such surface features as ocean swell, forests, or buildings, which all require energy to overcome. Along the same lines, water vapor convergence onto the continental atmosphere obviously requires water vapor, which is liquid water plus thermal energy; more required energy. Biogeochemical cycles too consume energy. First, they mostly revolve around photosynthesis, which consumes incoming solar radiation. They also consume energy for atmospheric and oceanic vertical mixing against stable stratification—for example, associated with lifting cold, dense mid-depth ocean water to near surface depths, where ambient density is lower. Additionally, most biogeochemical cycles include solid earth components that require, as discussed shortly, mantle motions against frictional resistance. Biogeochemical cycles, in short, consume energy.

Let's therefore look at the energy sources and motive forces of a few key fluxes, agricultural and otherwise.

The energy source for the hydrological cycle is the sun and its geographical variability. The most familiar observation here is that the tropics receive more incoming solar energy than high latitudes, annually averaging to roughly 400–450 and 150–200 watts per square meter at the top of the atmosphere, respectively.[415] Wide geographical disparities also describe energy absorbed by the surface, after atmospheric gases and particles reflected, scattered, and absorbed some incoming solar radiation, attenuating it, and after the surface reflected some of what's left. Earth's surface oceans absorb 25% more solar energy than land surfaces do on average, 165–175 vs 130–140 watts per square meter, but 2–3 *times* more of that absorbed energy goes into evaporating liquid water, 90–110 watts per square meter over Earth's global ocean, but under 35–45 over land.[639] And when more absorbed incoming solar energy is used for evaporation, less is left to warm the surface. For example, even if an incoming solar flux of 1,200 watts per square meter—vigorous even by tropical ocean standards—is fully absorbed by the surface ocean, evaporating about 1.8 mm per hour would completely undo this massive energy addition under typical conditions, keeping surface temperatures steady. These differences yield warming of the overlaying atmosphere by the surface at roughly double the rate over land than over the ocean.

These land–ocean disparate surface warming rates are greatly amplified by corresponding differences in three surface physical attributes that jointly determine how fast the surface warms under a given energy input.

The first attribute is the thickness h of the thermally active layer, the layer that discernibly responds to diurnal or seasonal variability in temperature and incoming solar radiation flux. In much of the ocean, that layer roughly coincides with the mixed layer and with the (minimally distinct) depth range within which temperature is essentially depth invariant, 20–50 m in warmer tropical oceans to 200–300 and beyond in rough, cold oceans.[301] On land, in sharp contrast, it is just a few meters. If you think about it, you kind of know this intuitively, because most basements are roughly 55°F (13°C) throughout the day and year. This means that their typical 3–4 m depth range already exceeds the depth of the thermally responsive layer over land, which is typically 1–2 m.

The second attribute is density ρ of the surface layer. In the upper ocean, it varies minimally within 1,020–1,030 kg per cubic meter. But over land, where proportions of sand, various rocks, or water vary widely, density is correspondingly variable, spanning roughly 1,500–3,000 kg per cubic meter.

The third attribute is a less familiar thermodynamic material property called specific heat, c_p, a measure of the energy needed to warm a kg of a given material by 1°C at a given fixed pressure (and surface pressure is, in almost all cases, no more than 20-30 millibars away from 1000 millibars, i.e., effectively constant). For rocks, soils, and water, c_p spans 200–600, 500–2,000, and about 4,200 joules per kg per degree (K), respectively.

Combining multiplicatively all three attributes, hourly warming by an absorbed energy flux of Q watts—or joules per second—per square meter is $3,600\,Q/(h\rho c_p)$. To illustrate the impact of the above surface properties on this warming, let's consider the surface warming rate under an absorbed solar flux of 500 watts per square meter, not atypical for daytime mean of clear summer days in the midlatitudes. Over the ocean, this would induce a surface warming of 0.1°–0.2°C per day. On land, in sharp contrast, it spans 0.1°–2°C per *hour*. With these warming rates, a 2°C warming (from 65°F to roughly 68°F, say) may take the ocean ten to twenty days, but one to twenty hours on land.

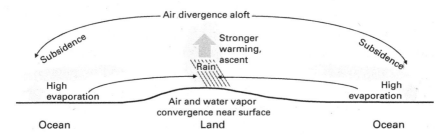

10.1 A schematic representation of a large scale land–ocean–atmosphere circulation pattern that arises, fundamentally, from the land–ocean thermal disparities and the obvious vastly higher availability of water to evaporate over the oceans than over most land masses. The text provides further details.

All of the above land–ocean disparities jointly impact warming of the lowermost atmosphere by the underlying surface, which happens twice as fast over land as over the ocean on average.[639] This is presented schematically in figure 10.1. With near surface continental air warming so much faster than oceanic air, the atmosphere over land is buoyed (e.g., at 10°C and 30°C, typical surface air density is about 1,240 and 1,150 g per cubic meter), thus ascending. Through ascent-related cooling by expansion, this makes condensation, cloud formation, and rain likelier. To supply this mean ascent over land, near-surface air must converge there (see converging arrows under the rainy area over land in figure 10.1, in which a continent is sandwiched between two oceans, like Africa with the Atlantic and Indian oceans to its west and east, and where air and water vapor fluxes are given by solid black lines and gray block arrows, respectively). Conversely, with subsidence over the oceans closing the circuit (left and right in figure 10.1), near surface air must diverge over the ocean (landward arrows). This near surface atmospheric divergence over the oceans combines with the much higher evaporation there to establish air and water vapor mass divergence over the surface ocean, especially in the tropics. The divergent lower atmosphere flow over the ocean is balanced by lower atmosphere convergence over land. This resupplies the lowermost continental atmosphere with water vapor—water vapor convergence is indeed robustly observed over all continents except Australia[662]—replenishing the vapor deficit due to the elevated rain propensity and closing the global water vapor budget.

Earth's carbon and nitrogen cycles are driven by several energy sources. Both cycles comprise fast and slow parts, related mostly to life and the solid earth, and operating on timescales of hourly to several years and millennia and beyond, respectively. The fast cycles are dominated by photosynthetic production of biomass and its subsequent decomposition, and are thus energetically driven by the sun. But both cycles also have abiotic parts. For example, one part of the carbon cycle arises from greater CO_2 solubility in cold seawater, about 3.5 g gas per kg seawater at 0°C but only 1.2 g per kg at 30°C. This, plus equatorial ocean dynamics not covered here, drive a long term mean carbon flux from the warm tropical oceans into the atmosphere, and carbon uptake by cold ones, with strong variability due to the coherent equatorial pacific variability known as El Niño, which results in large interannual variability in sea surface temperature throughout much of the central and eastern equatorial Pacific, with ocean outgassing there during warm years and uptake during cold ones. Another example is the much higher mean production of reactive nitrogen by lightnings over tropical land masses than almost anywhere else, especially polar regions, where almost none occurs.[525] In a lifeless but watery hypothetical Earth, this would have driven a poleward reactive nitrogen flux from the tropics. Because lightnings discharge electrostatic polarity, the *direct* energy source of this tropically concentrated reactive nitrogen production is atmospheric electricity. But none of this would have been possible in the absence of water vapor, which is to say, without evaporation from the tropical ocean. This too is therefore ultimately energetically driven by the sun.

An altogether *different energy source propels the agriculturally critical slow biogeochemical and soil cycles.* As mentioned above, slow biogeochemical cycles involve material exchanges between Earth's very thin crust on and in which we, and agriculture, dwell, and the very deep, hot mantle underneath. While the mantle is made of solid rock, over millions of years, it flows like a very viscous fluid. (If this strains your imagination, picture glass panes in historic churches, whose tops are considerably thinner than their bottoms, because glass too flows over centuries.) Most importantly, like a vigorously boiling pot of water whose surface is an ever-shifting pattern of "hills" and "dimples," the mantle also convects. Convecting fluids undergo vigorous vertical mixing by transient irregular rising and descending plumes that instantaneously underlay those "hills" and "dimples." This flow pattern arises when temperatures vary horizontally enough for some hot fluid patches to become sufficiently more buoyant than neighboring cooler ones to ascend, with fluid in the cooler areas descending. Mantle convection combines with gravity to drive plate tectonics, the process that constantly *very* slowly redraws Earth's map by geographically redistributing continents.

Two related surface expressions of plate tectonics play decisive roles in slow biogeochemical cycles: mid-ocean ridges, and subduction zones. The South Atlantic offers a perfect example of the former. Tracking the intriguingly mutually parallel coastlines of western Africa and eastern South America, parallelism that played a key role in the early history of the plate tectonics idea,[83,467] is an imposing underwater mountain chain. In some places, mid-ocean ridges rise above the ocean surface, forming islands (e.g., the Azores, Iceland). But mostly, they lay 2–3 km below the sea surface. While this sounds deep, it is some 2–3 km above the 4–5 km deep abyssal plain background. At the summit of this mountain chain, coherent mantle upwelling has managed to puncture the crust above, allowing new ocean floor to form from basalt that melts out of the upwelling mantle rocks. Because the whole mid–ocean ridge area is hundreds of degrees hotter than abyssal plains,[121] the mantle under the ridge is less dense than the cold mantle rock of which these plains are made (e.g., basalt density is 3,300 vs 3,180 kg per cubic meter at 300 and 1,600 K, roughly 80° and 2,400°F[489]). Thermal expansion (and thus lower density) of the hot mantle underlaying mid-ocean ridges explains the height of the ridge above the cold, dense abyssal plains. The fact that the ridge stands over 2 km above the neighboring abyssal plain creates a gravitational force that pushes the oceanic plates on either side of the ridge away from the ridge and each other in a direction perpendicular to the ridge axis.

The birthplace of oceanic plates is therefore the mid-ocean ridge. Being a geologic feature, their lifespan is long, mostly 100–200 million years, with rock age increasing with distance from the mid-ocean ridge (where it is zero). With this aging (which means longer cooling times since forming from erupted ridge basalt), plate temperatures correspondingly drop with distance from the ridge. As the oceanic plate matures (or, equivalently, as we sample its rock further and further from the ridge today), therefore, it cools, becomes denser, and sinks as it gradually morphs into deep, cold abyssal plain.

Thus gravitationally set in motion, Earth's surface plates move about and encounter one another. When two are thrusted toward each other, the less dense overrides the other, which subducts under the lighter one. The inevitable density differences that permit this can reflect differing histories, with younger, warmer oceanic plates being less dense than older, colder ones. They are even more pronounced when the colliding pair includes an oceanic and a continental one, because continental plates' density is 5–10% lower than oceanic plates'. Regardless of the origins of the density differences, and weakly related to their magnitude, as the two plates collide, the denser one slides under the more buoyant plate, bends toward Earth's center (typically by about 35°–55°), and begins sinking—or subducting—back into the mantle. This is the situation, for example, along the northwest coast of North America, where dense basaltic oceanic plates are thrust toward and subduct under the continental North American plate. Since subduction comprises denser objects sinking in less dense surroundings, it is gravitationally favorable, yielding another force that pulls the oceanic plate away from the mid-ocean ridge (the first, recall, was gravitational push by that topographically imposing ridge).

Over their life, oceanic plates thus journey from their mid-ocean ridge birthplace to their subduction zone graves under the oceanic water column. Over this 100–200-million-year journey, their upper surface, the ocean bottom, collects fallen remains of life above, especially algal primary production in the sunlit uppermost ocean, typically processed by higher trophic levels (animals that eat algae or algal remains and excrete their remains). Because some of this debris—"marine snow"—is organic matter, the very small fraction of it that survives oxidation (remineralization by rotting) contains such key elements of life as carbon, nitrogen, or phosphorus.[234] The mineral inorganic remainder, mostly shells, is likewise rich in other elements, notably calcium and magnesium. At the subduction zone, some of the material delivered with the oceanic plate "scrapes off" and builds such minor mountains as the California Coast Range, but much of it subducts. Much of the subducted mass volatilizes under the increasing temperatures descending plates experience deeper into the upper mantle, and returns to the atmosphere as volcanic gases. The remainder of this elemental bounty, however, remains with the subducting plate as it descends ever deeper into the mantle, eventually (after reaching depths of about 700 km) assimilating into the bulk mantle mixture. This completes the delivery of crustal material, derived by life or weathering of continental rock and soil, into the mantle. Because crust and mantle masses alike are approximately fixed in time on the mature Earth, over roughly 100–200 million years, an opposite mantle-to-crust flux must exist—in mid-ocean ridges and more exotic ascending mantle features—with magnitude that matches the subducting flux, a deep time biogeochemical cycle.

Since we just mentioned soil contributions to the land-to-sea-to-mantle flux, because soil is so central in agriculture, it makes sense to introduce the return flux that closes the soil cycle by generating the mineral component—essentially weathered crushed rock fragments—of new soils. This flux too is driven by plate tectonics, because rock weathering is faster on steep, young mountains only recently uplifted by tectonic forces.[334,641] Such mountain building—orogeny—follows collision of a continental plate with either another one (for which the rise of the Himalayas through the collision of what is now India with the bulk of the Asian continent is a perfect example), or on the landward side of oceanic subduction zones (exemplified by the Pacific coasts of the Americas, where such major mountain ranges as the Sierras or the Andes rise). Rivers then carry this enhanced sediment flux from the young mountains to sedimentary basins, coastal deltas, and eventually the deep sea, where it joins the plate tectonic cycle described above.

We thus explored several agriculturally critical cycles that consist of dynamic exchanges among various Earth reservoirs. We now need to introduce the motive force for these exchanges and cycles, the energy source for lifting the mid-ocean ridges 2–3 km above the abyssal plains, for volatilizing subducting material into volcanic gases, and for building mountain ranges on land. This energy arises from a very small internal energy source Earth possesses. This may seem puzzling, because the basic difference between stars and planets is the respective presence and absence of a significant heat source. Note the word "significant." Stars possess a prodigious internal energy source that fuels their bright luminosity.

For example, the sun, by no means a unique or unusually large star, shines about 4×10^{26}—4 with 26 zeros to its right—watts. Every second, in other words, it releases that many joules in all directions as, well, solar radiation. That internal energy source is nuclear fusion, mostly of four hydrogen nuclei, four protons, into an ever so slightly lighter (than its four constituent protons) helium nucleus with that small missing mass—the "mass defect"—liberated as energy. Too small for their gravity to press nuclei close enough together to permit fusion, planets do not have this energy source, and their energy balance is thus *very nearly* the balance between an incoming radiative flux from the star they orbit, and an almost equal radiative cooling to space from the top of their atmosphere. At the top of Earth's atmosphere, for example, the incoming flux from the sun is about 2×10^{17} watts. But Earth, like all not-too-old planets, also has a minuscule internal energy source, radioactive decay. (The relative youth condition reflects the fact that planets coalesce with their lifetime supply of radioactive material, and gradually go through it as they age, until there is none left, at which time the internal heat source dies down, as does plate tectonics. This is what I meant by "spent" planets early in the chapter.) Crudely and only imperfectly, radioactive decay is the opposite of fusion in that in it, larger unstable nuclei spontaneously break down and split into lighter daughter products. Radioactive potassium, uranium, and thorium are likely the dominant sources of heat in the Earth's interior.

Keep firmly in mind that Earth's radioactivity-based internal heat sources—about 50 terawatts or about 5×10^{13} joules per second—is vastly smaller than, about 0.03% of, the energy flux the sun regales Earth with, about 170,000 terawatts. It is even more trivial by comparison to stellar power output due to internal nuclear fusion. For example, while the sun outputs about 190 microwatts per kg of its mass, Earth's corresponding power output is only about 8×10^{-6} microwatts per kg of *its* mass, some eight orders of magnitude (one hundred million times) smaller. But diminutive though it is, over geologic timescales, it adds up to enough energy to drive all the above processes and thus hugely impact Earth's behavior and the nature of agriculture on its surface.

Because, as the preceding discussion makes clear, fluxes between reservoirs are the essence of the workings of Earth as a whole and agriculture within it, it would help to have a general tool that permits quantification and conceptual analysis of the functioning of these systems under widely ranging conditions. Let's refer to this structure as *the pool–flux model*, and use it as our guide for analyzing agriculture and its environmental impacts, for two reasons. First, agriculture is an integral part of Earth workings, as the entirety of this book argues and demonstrates, and attributes and substances in the Earth system are locked in never-ending cycles, constantly fluidly moving through Earth's various pools. The pool–flux model is therefore a most natural framework for examining Earth workings and agriculture's place in them.[10,354,461] Second, the pool–flux model rests directly on the bedrock foundational equations of predictive science, conservation equations, a suite that includes such matchlessly heavy hitters as Newton's laws of motion or the first law of thermodynamics. "Conservation equations" is a needlessly intimidating name, because at their core, such equations are intuitively obvious: what comes in must come out, and if not, the balance—"the in minus the out"—propels a steady buildup in the reservoir or pool in question.

A tad more formally, the nitrogen balance of an agricultural field over a cropping year can be cast as

$$\begin{pmatrix} \text{nitrogen} \\ \text{inventory on} \\ \text{December 31st} \end{pmatrix} - \begin{pmatrix} \text{nitrogen} \\ \text{inventory on} \\ \text{January 1st} \end{pmatrix}$$

$$= \begin{pmatrix} \text{sum of nitrogen} \\ \text{influxes} \\ \text{over the year} \end{pmatrix} - \begin{pmatrix} \text{sum of nitrogen} \\ \text{outfluxes} \\ \text{over the year} \end{pmatrix}$$

or, equivalently,

$$\begin{pmatrix} \text{annual nitrogen} \\ \text{accumulation} \end{pmatrix} = \begin{pmatrix} \text{sum of net nitrogen} \\ \text{influxes over the year} \end{pmatrix}.$$

Intuitive, isn't it?

To get rates or fluxes, "something per unit time," let's divide through by the one year duration,

$$\frac{\begin{pmatrix} \text{nitrogen inventory,} \\ \text{December 31st} \end{pmatrix} - \begin{pmatrix} \text{nitrogen inventory,} \\ \text{January 1st} \end{pmatrix}}{365 \text{ days a year}}$$

$$= \frac{\begin{pmatrix} \text{sum of net nitrogen influxes} \\ \text{over the year} \end{pmatrix}}{365 \text{ days a year}}$$

which becomes

$$\begin{pmatrix} \text{annual mean daily} \\ \text{nitrogen buildup rate} \end{pmatrix} = \begin{pmatrix} \text{annual mean net} \\ \text{daily nitrogen influx} \end{pmatrix}.$$

All of this seems like a pedantic expression of a fairly obvious idea: stuff (in this case nitrogen) neither appears nor disappears out of nowhere; only a net influx of stuff can augment the inventory. But I deliberately chose an obvious case to illustrate the point. In real physical systems, with many processes going on at once, changing many variables simultaneously, the simplicity of conservation equations become indispensable for keeping track of everything: fluxes in and out, accumulation rates, and so on. Conservation equations are the foundation of our knowledge of the ocean, atmosphere, forests, lakes, and indeed most physical systems,

To give you just one example, more current than historical, consider modern weather forecasting. The basic tool is a computer ("numerical") model of the atmosphere. All available global observations from surface stations, weather balloons, satellites, and various other sources define— after some very complex but not fundamental massaging—the state of the climate system at the initial time from which a simulation begins. From that point, the model marches forward in time in discrete leaps ("time steps"). At each such step, the future values of about fifty to one hundred state variables (generalizing the above single one, e.g., temperature, specific humidity) in about 0.2–1 million geographical locations ("grid points") and thirty to eighty heights in the atmosphere, are estimated. At each time step, therefore, several to hundreds of millions of equations are solved. Worse, in general, the equations for each variable at each point and height are coupled to *all* other equations at *all* other locations, and typically also not only currently, but also at several earlier times. In other words, all of those millions of conservation equations are *coupled*, very much like the two equations in two unknowns are in the culminating crown jewel of American high school algebra. And there is more; to get robust forecasts, tens of such solutions are obtained, each starting from a slightly randomly different climate state, in recognition of the inherent uncertainty that plagues our knowledge of climate states (e.g., can we really say definitively that today's temperature in Central Park is 29°C, not 28.99°C or 29.01°C?) And, to top things off, such conservation equation calculations are repeated in each of the world's weather forecasting centers several times a day, every single day. So yes, conservation equations are undoubtedly indispensable.

The other primary reason for the utility of the pool–flux model to our inquiry is that its structure captures perfectly (and is guided by) the cyclic nature of Earth's attributes—such as, energy, momentum, or angular momentum—and material substances we have already encountered. Material cycles—like the hydrological or the carbon cycles, or more narrowly, CO_2 uptake by plants, methane burping by sheep, or potash mining for fertilizer—are well known and were discussed earlier. I did mention in passing attribute exchanges earlier, but they are probably less familiar to most, yet intuitive once pointed out. A key example is slowing of surface winds by friction with underlying solid (land, forests, mountains) or liquid (lakes, oceans) surfaces. This slowing is the reason winds are stronger near mountaintops (think of Mt. Washington's record winds), especially isolated ones, than they are near sea level. This slowing reflects a downward momentum flux from the atmosphere to the solid or liquid surface, which is how a storm whips up huge ocean waves. It is also the same physical process that makes a car going at a fixed speed still consume energy to overcome wind resistance, even though by Newton's second law, cruising at fixed speeds should require no force. Because momentum is conserved (except for irreversible frictional losses into trivial amounts of heat), whatever the atmosphere loses in this interaction, the underlaying surface gains. While we just mentioned surface waves, this is particularly important for whole ocean basins. On those large scales, the momentum the ocean gains from the atmosphere via this flux is one of the principal forces driving upper ocean currents (the others headed by astronomical tides and geographical seawater density differences). This flux drives, for example, the subtropical gyres, the dynamical system whose western

North Atlantic expression—the Gulf Stream—is its best known feature. This flux also plays a key role in the El Niño phenomenon in the Equatorial Pacific. On smaller scales, this flux drives coastal upwelling of cold subsurface ocean water to the surface, the physics at fault for the thermally trying nature of swimming in La Jolla beach in summer under northerly winds.

Another important attribute exchange involves infrared radiation. When it's from the surface upward, this energy flux cools the surface and warms the atmosphere. An opposite energy flux also exists: downward radiation from atmospheric gas molecules and particles (notably cloud water droplets or ice particles) that warms the surface and cools the atmospheric level from which it originates. And a more exotic attribute flux: big northern hemisphere midlatitude storms apply enough eastward torque on such mountain ranges as the Rockies or the Alps to actually speed up a bit Earth's eastward rotation, thereby shortening the day by a few milliseconds. (The effect dies down, returning rotation to its normal rate, once the storm ends.) In yet another important example, the atmosphere and land always exchange heat by additional means beyond infrared radiation. For example, imagine parking near the beach on a hot summer day, and walking barefoot toward the beach. The grassy parking lot is cooler, but the sand is much hotter. Familiar? There are several simple reasons for this, but the key one is evaporation. Depending where and when, grass loses 0.3–1 mm per hour to total evaporation (through the leaves plus directly from the soil) on a typical summer day, or 0.3–1 liters per hour per square meter. Noting that a liter of water weighs 1 kg, a squared meter of grass evaporates about 0.3–1 kg water per hour. Since evaporating a kilogram of water requires roughly 2.4 million joules of energy, which in this case is extracted straight from the surface (the grass and underlying soil), the above daytime evaporation rate cools the grass at a rate of 200–700 watts per square meter (the last step uses the watt = joule per second equivalence). To get a sense for the meaning of this cooling rate, let's compare it to natural summer daytime rates of heating by the sun, typically somewhere inside 650–1,100 watts per square meter. The 200–700 watts per square meter cooling rates we calculated above for the grass thus amount to undoing one fifth of to *all* surface warming by the sun! Evaporative cooling can therefore dominate total surface heat flux even on land, especially agricultural land, and often does. Because agriculture typically enhances markedly evaporation rates, and our dietary choices impact our agricultural land needs, surface temperatures and the vigor of heat and water vapor surface–atmosphere exchanges bear the unambiguous signature of these choices. [143,179,508,598] Add the primary import of the surface as a source of heat and water vapor to the atmosphere over land, and the fundamental importance of energy, water vapor, carbon, and nitrogen fluxes to the behavior of various Earth pools (notably near surface land environments, where most Earth life resides) becomes abundantly clear in natural and agricultural areas alike. This is what makes the pool–flux model so helpful and informative.

Resting on the solid foundation of conservation of mass, energy, and momentum, the pool–flux model is widely used for conceptualization and qualitative understanding of physical systems, and for predictions. Because of the ubiquity and import of this use, let's see briefly how conservation statements can be applied for predictions.

Imagine we wish to study phosphorus dynamics in a small pond. Let's denote the phosphorus inventory in the pond as time t by P_t: for example, P_{Monday} gives the number of kilograms of phosphorus the pond contains on Monday. Let's suppose the pond is fed by one stream and is drained by another, whose respective phosphorus fluxes are $f_{in\,or\,out}^{day}$, e.g., $f_{in}^{Tuesday}$ gives the number of kg of phosphorus the streams delivered into the pond on a given Tuesday such that the net phosphorus influx is $f_{net}^{Tuesday} = f_{in}^{Tuesday} - f_{out}^{Tuesday}$. Suppose we measure these fluxes on Monday, Tuesday, and Wednesday, thus deriving a representative mean net flux in kilograms of phosphorus per day,

$$f_{net}^{mean} = \frac{1}{3} \left[\left(f_{in}^{Monday} - f_{out}^{Monday} \right) + \left(f_{in}^{Tuesday} - f_{out}^{Tuesday} \right) \right.$$
$$\left. + \left(f_{in}^{Wednesday} - f_{out}^{Wednesday} \right) \right].$$

Using measured Monday and subsequent Wednesday inventories, and assuming the right-hand side (the net influx) varies tolerably over a few days, phosphorus conservation states that

$$P_{\text{Wednesday}} - P_{\text{Monday}} = f_{\text{net}}^{\text{mean}} \times 2 \text{ days}.$$

The change in the pond's phosphorus content in kilograms (on the left) therefore equals the net phosphorus flux into the pond, over the Monday-to-Wednesday two full day duration. Rearranged, the above reads

$$P_{\text{Wednesday}} = P_{\text{Monday}} + 2f_{\text{net}}^{\text{mean}},$$

in which all terms on the right are known, so that the forecast, the left-hand side, can be calculated. This is a highly simplified example of the use of conservation statements for prediction, a prognostic calculation in scientific lingo. At least in principle, if we know the pond's phosphorus content on Monday, and have a decent estimate of the net flux in (the right-hand terms of the equation, respectively), we can prognosticate about the pond's phosphorus content on the following Wednesday. Because the same basic structure applies to masses of carbon, water, or nitrogen, and to fields, lakes, the atmosphere, or oceans, the pool–flux model is a fundamentally useful tool not only for studying natural systems but also for making predictions about their future states.

One practically important use of the pool–flux framework is obtaining a rough estimate of the importance of a given anthropogenic flux. Because for the most part agriculture modifies natural fluxes rather than introduce altogether new ones, one way to quantify this is to express the agricultural flux into or out of a given reservoir, call it f_{ag}, as percentage of the corresponding natural fluxes. One way of doing so is

$$f_{\text{relative}} := 100 \frac{f_{\text{ag}}}{f_{\text{net}}^{\text{natural}}}$$

where the denominator is a defensible characterization of the natural fluxes into or out of the pool in question. When f_{relative} is small, say 1–2%, the agricultural modification is small relative to and likely dominated by natural fluxes. The reverse—$f_{\text{relative}} = 1{,}000\%$, $10{,}000\%$, and beyond—describes such environment perversions as an intensive 800-head dairy herd draining into a single small creek. While the creek surely has natural nutrient sources, like every natural system, those are dwarfed manyfold by the dairy farm's effluent. Another measure of the importance of an anthropogenic flux is its size relative to that of the pool it is connected with, which gives rise to a timescale. Denoting the characteristic reservoir size R, our timescale (denoted by the lowercase Greek tau) is

$$\tau := \frac{R}{f_{net}^{mean}}.$$

Notice the units: something divided by the same something per unit time (e.g., R in kg nitrogen per kg of river water divided by f_{net}^{mean} in kg nitrogen per kg of river water per day), yielding a ratio (the timescale τ) in the same time units used to quantify the mean net flux (days in the example). Below, I illustrate the utility of both measures by using them to quantify the importance of anthropogenic greenhouse gas emissions to the totality of the anthropogenic climate change problem.

In more complex system, a key new variant of the same principles arises, yielding one of the most versatile and widely applicable tools in the natural sciences, the numerical model, which we introduced in our earlier discussion of climate modeling or weather forecasting. Numerical models typically arise from dividing any large object of inquiry—the atmosphere, say, or the Indian Ocean—into many smaller imaginary sub-objects, or control volumes, and applying the pool–flux model to each individually, but in a framework that recognizes their mutual connectivity. The reason this variant is so important is that these control volumes allow the right hand flux terms to also be internally calculated by the model, by simply judiciously applying conservation principles to each substance inside each control volume at each time step as the model marches forward toward tomorrow, the next day, and beyond.

If this was somewhat cryptic, imagine dividing the atmosphere into such small virtual boxes, and focusing on the one spanning 38°–40°N, 79°–81°W, and 3–3.2 km above the local surface. Let's imagine that around 9:10 a.m., there happens to be no vertical (upward) or meridional (northward) wind. The only nonzero wind in the vicinity of the above delineated imaginary box is eastward, denoted u (westward wind is simply negative u). The net water vapor flux into our box at 9:10 a.m. is then

$$f_{q,net}^{09:10a.m.,39°N,80°W,3.1\ km} = A\rho_{air}\Big[(uq)^{09:10a.m.,39°N,81°W,3.1km}$$

$$-(uq)^{09:10a.m.,39°N,79°W,3.1km}\Big].$$

Preceding the brackets is A, the area of the east- and west-facing faces of our control box in square meters, times ρ_{air}, air density in kg per cubic meter. Within the brackets, q denotes the air's specific humidity in g water vapor per kg air, and the superscript defines the location where we measure both the eastward wind u and the specific humidity q. Jointly, this yields a flux, g water vapor per second. The first measurement is at 81°W, the west face of our box, midway between its center and that of the box immediately to its west. (Box centers are where all this accounting refers to; what happens between neighboring box centers is formally unknown, but one can sort of "fill in the blanks" formally or informally.) Consistently, the second measurement is at 79°W, the east face of our box, midway between its center and that of the box immediately to its east. With that, the terms inside the brackets are water vapor fluxes into and out of our box by the eastward wind. The difference, influx minus outflux, is the net influx, the left-hand term. The latitude and height of the measurements correspond to the center of our box, because there are no upward or northward winds, so we only address transport by eastward flows, and for those, west is upstream and east is downstream, defining the locations of the in- and outfluxes of our control volume. When they do exist, northward or upward flows are easily handled in corresponding ways. And with this, we magically have a way of calculating the net flux right-hand term from information we already have, and that updates anew for each time step. This is what I meant above by the right-hand terms being "also internally calculated by the model." But there is a price to pay, the need for one additional class of information, boundary conditions. In our example, it would mean the value the westward box would have had just west of a western boundary, if it existed. For example, for a model of the North Atlantic ocean, what do we do near the east coast of the US, where further west is only land? We need boundary conditions. For temperature, for example, one reasonable choice is using nearshore water temperatures measured along the eastern seaboard.

This structure is applicable to any object of interest, in any natural or managed system around us. Astonishing though it may sound, this application of generalized conservation equations—of the *pool–flux model*—is the essence of a huge swath of predictive physical science. To forecast the weather two days hence, or the position of Mercury a year from now, or

Earth's climate in 2075, far more complex renditions of essentially the above very principles are used, based on Newton's second law of motion, the first law of thermodynamics, among other conservation equations.

The pool–flux model is thus extremely robust, widely applicable, and scale general to objects of inquiry as small as a steer or a field or as large as an agricultural region or a continent. Used as the prism through which food production is examined, the model highlights a fact that is rather obvious yet often ignored, that agriculture is a local, sometimes minor, and in other instances dominant subset of the great planetary cycles with which it is inextricably coupled. This is why stripping agriculture of its geophysical–planetary context, viewing it instead as an isolated, localized industrial process to be optimized using industrial engineering tools is environmentally ruinous, yet sadly also the foundational bedrock of industrial agriculture. To be clear, there is nothing wrong with optimizing agriculture per unit input, which is almost impossible to rationally oppose. What *is* wrong, as the pool–flux view so clearly illuminates, is optimizing agriculture with complete disregard to basic immutable constraints Earth imposes on it, as though agriculture somehow enjoys a waiver from Earth's basic working principles. (A literal such waiver for agriculture in the US Clean Water act is a tragic example of this irrational, ruinous yet pervasive view of agriculture.) Let's therefore return to the usage of the pool–flux model for conceptual exploration of agriculture as a subset of Earth functioning,[253,332] alternating our focus between Earth and agriculture.

The pool–flux model helps placing irrigation back on its hydrological footing. Modern irrigated agricultural provinces like the Yizrael Valley of northern Israel or the Central Valley of California, typically derive their water from afar. This can mean relying on regional or national water redistribution systems, or on tapping aquifers (water naturally stored underground). Either way, these redistributions amount to taking from where there is excess (from precipitation, dew, or fog influx[144,260,524]) to where there isn't enough.[378] Involving reservoirs and requiring fluxes, these redistributions are thus clear examples of what I described earlier as "concentrating more ... of some things ... for enhancing useful removal," adding water to a field or an agricultural region and enhancing yields, respectively. The pool–flux model is therefore the natural analytic tool for quantifying the environmental impacts of irrigation.

Given this book's environmental focus, you may reasonably wonder: if irrigation has to be analyzed, using the pool–flux model or otherwise, an environmental problem lurks somewhere here. But what's the matter with irrigation as depicted above? More broadly, what's wrong with greatly speeding up a given flux in a pool–flux system? After all, most any place on earth naturally receives at least some water input, so irrigation changes the system only quantitatively, not qualitatively. The answer is that because agriculture is embedded in larger planetary cycles, its impacts likely propagate throughout much of the Earth system, where they are likely to have unintended consequences, some adverse. An example we have already encountered involves reduced local surface temperatures and increased humidity and precipitation by irrigation both locally and downstream.[560,635] By enriching the overlaying air with water vapor, irrigation also reduces the diurnal temperature range, raising nighttime minima while reducing daytime maxima.[432] Locally, the effects can become truly gargantuan. For example, in the heavily irrigated Central Valley of California, the greatly enhanced availability of liquid surface water to be evaporated combines with the very dry, well ventilated boundary layer to raise markedly summer surface evaporative cooling, by some estimates reaching 4°C (7°F)[112]; imagine the difference between a midday summer temperature of 83°F and 90°F (28°C vs 32°C). These effects all represent accelerated local- to continental-scale water recycling: faster surface evaporation, more condensation in the air above, raising cloud cover and precipitation. These changes can propagate throughout the hydrological cycle and the Earth system, including to areas and processes far afield and not directly perturbed by agriculture,[353] with magnitudes that are readily predictable by the pool–flux model.[11,127]

The pool–flux model is also useful for placing fertilization in its biogeochemical context. Think of the most common forms of fertilization, spreading of synthetic fertilizer, manure, or potash, or the less common but widely used in organic operations "green manuring" (see Accompanying PDF). Like irrigation, these and other fertilization methods raise the local cap on plant growth imposed by natural supplies through "concentrating more of some things." Like in the preceding irrigation case, there is nothing novel or "unnatural" here, just augmenting agroecosystems' nutrient supplies well beyond levels afforded by such natural influxes as atmospheric nitrogen deposition as dry particles or as solutes in precipitation, dissolution of underlying rock minerals, or influx of riverine solutes from upstream. You may object, because industrially produced synthetic fertilizer is obviously not natural. But is its production meaningfully distinct from nitrogen oxide production in hellishly hot air around lightning rods, where cooling plasma—matter so energetic that its electrons are free, unmoored to nuclei—yields lone nitrogen atoms that can bond with atmospheric oxygen, eventually becoming soil nitrate available for plants? Fertilization too therefore amounts to accelerating naturally occurring fluxes into agricultural productive units, and the potential troubles arise downstream, the most environmentally and societally devastating, best known, and mechanistically understood example being oceanic "dead zones."[55,212]

Dead zones are discussed extensively elsewhere,[273,675] but a brief retelling of this story is appropriate here. It starts with over-fertilization, applying more than crop plants can take up. The excess dissolves in drainage water, and flows toward the ocean in streams and rivers. Because river water is fresh, it is 2–3% less dense than typical seawater. At the river's mouth, buoyant river water floats on top of the salty coastal ocean, forming a thin fresh surface layer. Because this freshwater also delivers nutrients from the unused fertilizer to the ocean, the surface coastal ocean becomes nutrient rich. In summer, the combination of abundant solar radiation, and a warm, nutrient-rich surface layer is ideal for algal growth, promoting an algal bloom, or population explosion. After their brief life, algae die and their remains eventually sink to the bottom, where they— like a stray head of lettuce you left way back in the fridge and lost contact with—rot. But because rotting, or organic matter decomposition, is oxidation, it consumes oxygen. This means that much of the oxygen normally dissolved in near-bottom water is used up by all that excess rotting. This depresses levels of near-bottom dissolved oxygen, leaving too little for most oceanic life forms—for example, fish, shrimps, or my wife's favorite Chesapeake Bay crabs—that require oxygen. In extreme cases, which now occur entirely predictably (by our pool–flux model, say) every summer in the northern Gulf of Mexico, Chesapeake Bay, or Lake Erie, among other inland and coastal bodies of water, widespread death occurs. In dead zone research, the pool–flux model is extremely important both for predicting the magnitude of next summer's dead zone given expected fertilization information, and for using historical data on fertilization and dead zone severity and extent to estimate fertilization efficiency, the fraction of total applied fertilizer that ends up in usefully harvested crops.

With slight modifications, the same general points also apply to harvest. For example, consider an Idaho alfalfa field, from every acre of which 4–4.5 tons of hay are removed annually. At first blush, this seems utterly without natural precedents. First, such yields handily eclipse what this land would have naturally yielded even in the best of years. In addition, all of it—mowing, raking, baling, shipping to far away destinations—is mechanized and powered by fossil fuel, and none exists in nature. True. But did bison and wild horses not graze this field semi-regularly before the mid-1800s, or even the 1500s? Did mammoths, mastodons, and myriad other now extinct large herbivores not do the same in the Pleistocene before? And did these large herbivores not remove aboveground vegetation and carry it in their stomachs far away? So hay production is not qualitatively distinct from what naturally occurs, it is just a greatly accelerated and geographically stretched version of natural biomass removal and redistribution by secondary producers.

By this point, you surely recognize the accelerated flux in this story, harvested yield, or rate of biomass removal from the average Idaho alfalfa field. Implicitly, though, it is clear that this must be accompanied by commensurate augmentation of influxes of nutrients and water, and reduction of others (such as weed growth). And the local or downstream downsides? Greatly increased soil erosion (high yield relies heavily on mechanical cultivation, which promotes soil erosion),[46,406] declining soil organic carbon and nutrient concentration,[327] compromised soil microbiota,[387] among other adverse impacts of the deliberate enhancement of the yield flux.

For basic agricultural production units—a beef ranch, say—agriculture is thus indeed about concentrating more of some things than is naturally occurring, notably accelerating influxes of water and nutrients, and slowing other fluxes (weed growth in a field, beef loss to predators), all in the name of enhancing yield. While this wording may be a bit unorthodox, this view of agriculture is almost self evidently true, as best I can tell offering nothing to object to, even by the great Wendell Berry.[53] Most importantly, by explicitly mechanistically connecting agriculture with Earth's churning, this view replaces the traditional naively reductionist and misleading view of food production as just another industrial process in the countless many that make the modern world tick with a useful, workable one that places agriculture where it rightfully belongs, at the heart of human planetary agency. This replacement stems naturally from the observation that the very same structure and basic building blocks—reservoirs exchanging substances and attributes via fluxes—can perfectly well also serve as the closest thing to a universal description of environmental problems, as we will see in detail shortly. So agriculture and environmental integrity are close siblings, following extremely similar blueprints. And, as is not uncommon among close siblings, they can develop rivalries, some transient, others deeply entrenched.

To illustrate this duality, let's use the pool–flux model to interpret our most urgent and all-encompassing environmental problem, anthropogenic climate change, and its relations with relevant natural Earth processes. Because CO_2 is the main anthropogenic greenhouse gas, carbon and the atmosphere are the focal substance and pool, respectively. Combining the two, the root cause of anthropogenic climate change is of course a small atmospheric carbon imbalance arising from CO_2 removal from the atmosphere being slightly smaller than fluxes into it. Dominating the net atmospheric C imbalance is famously anthropogenic emissions, enhanced solid earth to atmosphere flux in the language of the pool–flux model. Currently, this imbalance is about 10 Pg carbon a year (1 Pg C $= 10^{12}$ kg carbon $= 1$ trillion or 1,000 billion kg carbon). How small is this? First, while many fluxes connect atmospheric carbon to carbon in other cogs of Earth's machinery, because these fluxes are decisively dominated by exchanges with the terrestrial biota and ocean, it would make sense to directly compare anthropogenic emissions to the roughly 200 Pg C the ocean and land biota jointly annually release—almost all due to respiration, the CO_2 released from organic matter rotting—into the atmosphere. Against this reference, 10 Pg C a year corresponds to an $f_{relative}$ of about 5%. Clearly, even a 95% perfect process can go terribly awry. Another relevant yardstick is the time it would take the anthropogenic carbon flux into the atmosphere to, arbitrarily, double atmospheric carbon. Dividing the atmospheric pool, currently about 830 Pg but sadly still rising, by the 10 Pg C a year anthropogenic flux yields $\tau \approx 83$ years. So as things currently stand, it will take atmospheric emissions roughly a century to double current atmospheric carbon mass. Beginning to appreciate just how basic fluxes, reservoirs, and their delicate tango are? Even a modest imbalance is enough to put us in grave environmental difficulty if it persists, and human carbon emissions appear nothing if not recalcitrantly persistent. The remainder of this chapter addresses a few dominant carbon fluxes between Earth reservoirs we know quantitatively courtesy of innovative applications of the pool–flux model by generations of geochemists and biological and chemical oceanographers.

Essentially the mirror image of the 200 Pg C a year respiration flux mentioned earlier is uptake of CO_2 by primary producers, partitioned as roughly 60:40 among land and ocean. In the ocean, atmospheric CO_2 invasion into the upper ocean is driven by uptake of CO_2 dissolved in seawater by algae photosynthesizing near the ocean surface. Because—courtesy of the second law of thermodynamics—stuff flows spontaneously down gradients, toward lower concentration, the elevated atmospheric CO_2 abundance combines with this removal from the upper ocean to provide the driving force underlying this "ocean invasion." But this can easily reverse when photosynthesis slows or stops. One way to strongly limit photosynthesis and the corresponding downward CO_2 flux into the upper ocean is to limit sunlight or make the water too cold (which is good for solubility, but when it becomes too extreme, it greatly slows down biochemistry, and with it the rate of photosynthetic CO_2 removal from the upper ocean by algae). Both too cold ocean surface *and* too little sunlight at once are typical of mid- and high latitude oceans (e.g., the US east coast north of Cape Cod, say) in fall and winter. I vividly remember this effect from the time I was a junior scientist at the Woods Hole Oceanographic Institution in the early 2000s. I'd conclude my bike ride to work with a morning swim at the pebbled beach right near the Ocean Physics lab, overlooking Martha's Vineyard. Throughout summer and into early fall, the Vineyard Sound water tended to be on the greenish murky side, teeming with algal life (i.e., productive). And then, one clear and chilly fall day, typically around mid-October, the water would become almost magically clear, revealing the bottom pebbles that exacted such a painful toll on my bare feet as I waded in during the preceding seven months. And by a strange coincidence, that day was also the first day that getting into the water also involved a briefly arrested breath, as the sun was receding to

SMOOTH EARTH OPERATIONS: RESERVOIRS, FLUXES, CYCLES

the south, the days were getting shorter, and the MIT fall semester was approaching midterms... Productivity can also be suppressed by limited availability of such nutrients as nitrogen or phosphorus that are required for biomass production. Because the ocean receives much of its nutrients from continental weathering, nutrients are more likely to limit productivity in oceanic interiors, far from the relative nutrient bounty of coastal waters. Quintessential examples of this type of nutrient limitation are so-called "oceanic deserts," such as the Sargasso Sea (so beautifully depicted in *Islands in the Stream*[257]), where surface winds collaborate with Earth's latitude dependent eastward rotational speed (a mouthful synonymous with "the Coriolis force") to compel surface waters to converge and downwell into the subsurface and beyond,[409,576] leaving the surface with old, nutrient-depleted waters that can sustain little primary production.

Regardless of the reasons for the reduced photosynthetic CO_2 uptake in the surface water, respiration—again, essentially the rotting of dead algal matter, call it "reverse photosynthesis"—continues, because it requires neither sunlight nor nutrients (it does require oxygen, whose dearth can slow respiration, but surface water O_2 concentrations low enough to stop rotting are naturally not very common). Because aerobic respiration is ubiquitous throughout much of the upper ocean, and it enriches the surface water with its end product, CO_2, when not outpaced by opposing photosynthetic CO_2 uptake, it can enrich the surface ocean of CO_2 enough to force upper-ocean degassing of carbon into the atmosphere, attenuating or even locally reversing "ocean invasion."

Expanding on the solubility considerations mentioned earlier, some ocean–atmosphere carbon fluxes arise thermodynamically, that is, independently of life and purely from basic physics of the gas–liquid system that the lower atmosphere and upper ocean jointly form. The key bit of physics governing these abiotic fluxes is the temperature dependence of CO_2 solubility. Recall that from 30°C to −1°C, which spans almost all observed seawater temperatures, solubility approximately triples, from 1.2 to 3.4 g CO_2 per kg ocean water, respectively. These variable solubilities too can impact the magnitude and direction of abiotic ocean–atmosphere CO_2 fluxes (recall the second law of thermodynamics).

On the atmosphere side, CO_2 concentrations vary minimally. The largest difference occurs in mid- to high northern latitudes between a

May maximum and a September minimum, mostly reflecting photosynthetic plant uptake in the preceding three to four months. To understand this requires keeping in mind that continental productivity is far larger than the ocean's, accounting for roughly 60% of global biotic uptake despite occupying only about 30% of the global surface area. Consequently, continental seasonality controls atmospheric CO_2 seasonality, and because in today's Earth most continents are huddled together by chance in the northern hemisphere, that hemisphere's seasonality controls atmospheric CO_2 seasonality. Because northern winter uptake is minimal, CO_2 gradually builds up between roughly November and May. Conversely, vigorous northern summer uptake gradually depresses CO_2, leading to the September minimum. But the amplitude of this (May maximum minus September minimum) is only 15–18 parts per million, 3–4% of recent mean atmospheric CO_2, about 415 parts per million. Further south, where the surface is mostly low productivity ocean, photosynthetic uptake and CO_2 seasonality are low, a mere 1–2 parts per million at the South Pole. For the most part, therefore, atmospheric CO_2 variability contributes much less to ocean–atmosphere abiotic CO_2 flux variability than the threefold temperature-dependent change in CO_2 solubility in seawater we saw earlier, along the line of the following observations.

An example of local abiotic ocean uptake of atmospheric CO_2 arises off the New England coast around March–April. Because the late winter surface seawater is not much above freezing, it can hold in solution 3.2–3.4 g CO_2 per kg ocean water. But being just north of the warm Gulf Stream, the surface ocean there is dominated by old, nutrient-exhausted water that was likely much warmer just recently, as Sargasso Sea water with much lower dissolved CO_2 than the maximum it can hold in its new low temperature. The surface ocean is therefore "CO_2 deficient," holding less than it can. CO_2 molecules just above the air–water interface are therefore far more likely to assimilate into the liquid than they would have been had the water been near CO_2 saturation. The result is a vigorous atmosphere-to-ocean CO_2 flux, a mostly thermodynamic ocean invasion that's only indirectly related to life.

Just like its biological counterpart, the thermodynamic flux too can reverse into ocean degassing. The quintessential example of this takes place in the eastern equatorial Pacific, whose ocean dynamics are in some crude way superficially roughly the opposite of Sargasso Sea dynamics, with wind driven divergence of surface water driving upwelling of nutrient-rich subsurface water to the surface. The upwelled subsurface water is CO_2 enriched by decay of dead algal remains that "rain" down into the subsurface ocean from the algae teeming surface. Because it originates at depths well below the reach of the bright tropical sun, the upwelled water is also much colder than the surface and thus able to dissolve much of that extra CO_2. But once it surfaces, the cold water warms rapidly by that bright sun, quickly depressing CO_2 solubility in it, forcing the equatorial Pacific ocean to "exhale" excess CO_2, degassing carbon into the atmosphere.

Combined, the biological and thermodynamic ocean–atmosphere carbon fluxes are absolutely central to climate and more broadly to Earth system functioning. They connect the atmosphere with the ocean and— because some carbon in ocean floor sediments eventually returns to the mantle in subduction zones—with the solid Earth. This fundamental import illustrates the centrality of fluxes and exchanges among reservoirs to Earth's normal functioning[16,17] and to environmental problems that arise when this functioning is disrupted.[10,354,461]

11

Bringing Planetary Eating Home: How to Shop, Cook, and Eat to Maximize Environmental and Nutritional Benefits

For some, it follows such unmistakable harbingers as your daughter informing you that her lunch comprised nothing more than instant ramen-in-a-cup and some salty dust from the bottom of what was once a proud bag of corn chips. For others, I hear with some incredulity, it may be a specific pre-assigned day of the week. Regardless of grocery shopping tribal affiliation, life eventually requires restocking. While many view this as just another one of life's interminable chores, I love grocery shopping. Not in such food boutiques as Whole Foods or the Berkshire Food Co-Op (although it *is* incredibly seductive to briefly surrender to the sticky embrace of the fantasies they sell), but in places that sell food: plain, mutually indistinguishable American chain supermarkets. I love them so much, in fact, that when traveling, unless there is a Lebanese or Syrian place nearby from which I can emerge happy and full for under $12 (circa 2023) in under twenty minutes, this is where I get all my meals.

I love grocery shopping so much because it matters. Greatly. For one thing, you can markedly reduce your greenhouse gas emissions. As introduced and reintroduced throughout the book, food accounts for about a quarter to a third of our greenhouse gas emissions,[133,605] a share somewhat smaller than our need for industrial goods,[276] but larger than both transportation (14%) and powering the built environment (18%). Moreover, variability among different food choices is so large, ten- to fiftyfold is not exceptional, that you can easily quadruple your one meal emissions with one bad choice, or cut it by 90% with one good one. Our diet is thus a very logical focus of efforts to reduce greenhouse gas emissions, and while governments could have played a key positive role in this, a long record shows that they mostly prefer not to.

But identifying the good choices requires solid guiding knowledge that minimizes wishful thinking[35,563] by building on rigorous quantitative analyses.[179,180,183,475,595] While I hope this book has helped you negotiate this foundational challenge, deploying it in practice—planning an actual meal, say—is its own formidable obstacle. Helping you scale this obstacle is the purpose of this final chapter. Some may question whether individual dietary modifications can be adopted widely enough to become meaningful. It is a reasonable, fair point, but one that also applies to buying EVs, installing rooftop solar panels, insulating drafty homes, or upgrading windows, all of which have been hounding the discredited traditional methods they replace at breakneck speeds. Moreover, while the above voluntary actions are guided only by benevolent care and promises of eventual financial returns, food choices also affect health, and authoritative expert advice on how to optimally combine the two goals is fortunately available (e.g., ref. 577 or less compromised alternatives[61,347,643]). It is thus reasonable to expect that if widely accepted environmentally motivated dietary guidelines are developed—for which the EAT-Lancet paper[642] is a useful preliminary step—public following of those guidelines will be expansive enough to produce wide overall impact.

And the benefits of thoughtful dietary choices go beyond just emissions. They also include ceding to biodiversity swaths of wild land[522] large enough to be visible from space (e.g., figures 3–4 in Hoang et al.[261]), and dramatically reduce water[183] or air pollution-related mortality[156] among people unlucky and legally vulnerable enough to neighbor careless farms.

With that backdrop in mind, grocery shopping can suddenly be elevated from mindless tedium to our chance to wrest control from frustrating systems of governance,[310,463] and make the impact of our choosing, at the timing and by the means of our choosing. Is the fantasy of such unfettered agency not the essence of the popularity of superhero movies among the post-eleven-year-old set?

BRINGING PLANETARY EATING HOME: HOW TO SHOP, COOK, AND EAT   201

Thus empowered, let's cross the automatic sliding glass doors and enter the only part of the American food system into which we are welcome.

In most supermarkets, your first stop is the fruit and vegetable section.[425] Most anywhere I travel, the produce aisle is quite lovely, bountiful, and diverse these days. This is *huge* progress; when I first came to the US, the American produce aisle specialized in wilted iceberg lettuce and waxy tomatoes whose flavors were indistinguishable both from each other and from the wet cardboard boxes from which they emerged. As I walked into the produce aisle of our local Tops earlier today, I noticed a nice looking fresh fennel, or, as I fondly remember it from biking across Italy, finocchio. Since this is somewhat unusual, let's make finocchio salad for lunch. Now I eat fennel every day, but mostly as dried seeds, not the fresh white-greenish plant. Fennel is great for breakfast: mine often consists of a cubed pink grapefruit, happily swimming in olive oil, with lots of fennel seeds, black sesame seeds, extra hot dried red chili peppers, fresh turmeric root, paprika, cardamon, and toasted *un*salted pumpkin seeds in the shell sprinkled on top. Because I only eat plants, I also sprinkle nutritional yeast on top, a fine B_{12} source for plant eaters. If you try (and hopefully like) this breakfast, please do *not* discard the whole thick grapefruit skin. Instead, cut off only the pinkish skin with a sharp knife, leaving as much of the white pith underneath as your knife skills permit. (And do *not* let my paper thin unbroken peel discourage you; it took a while to perfect...)

Beyond a great breakfast, this simple combination highlights several general, important issues.

First, why save the pith? A better question is, why discard it? If we have already incurred all the financial and environmental costs of growing something, is it not obviously better to use as much of it as possible, unless it is—like melon rind, say—genuinely useless? While not a major challenge—waste management generates 3–5% of all global greenhouse gas emissions[109,219]—emissions from organic refuse expands the scope of our task of environmentally soundly disposing of it. If you ever visited a municipal dump, you must have noticed those inverted U-shaped plastic pipes jutting upward, cane-like, out of inactive, earth covered piles. These are methane vents, preventing buildup inside the pile of this potentially explosive gas (the main ingredient in natural gas), which also happened to be a powerful greenhouse gas. They steadily leak the gas, which is the end product of anaerobic decomposition of organic matter in the pile, into the atmosphere, where it strongly if briefly worsens our climate challenge. This is why I eat every last bit of our broccoli stems (finely dicing microwaving, and dousing them with Dijon mustard and freshly squeezed lemon juice makes a real delicacy; give it a try). This is why I finely chop my lemon peels, after squeezing them for juice, and spread them on my salads, discarding nothing. And this is likewise why I eat roasted unsalted peanuts in their intact hard shells. A bit of decidedly uncomplicated chewing, and you treat yourself to a nutritious (see below) delicacy. My only "trail mix" for unusually long bike rides is a mixture of about three to four parts of these peanuts with one part diced dates or figs; this has propelled me up Mount Washington many a time... Don't interpret the above to mean that such odd customs are necessary; the problem they reduce is nowhere near the top of the list of our problems. But they do reduce it markedly, they enhance the flavors of our food, they redirect your money from processors to honest producers, and they promote health.

The pith's slight bitterness greatly enhances the flavor of the remaining breakfast elements, including the unmistakable taste of the sheikh of all food, olive oil. When I can find it, I buy Palestinian oil; nothing else comes close. The giveaway is the black sediment near the bottom of the undisturbed jug. If you are lucky enough to find such oil (exploding settlers' violence in the occupied West Bank has made it *very* challenging), the thing to do is mount a garlic clove on a long, pointy branch, dip it in the tar-like bottom residue, sweep it in the same 270° folding motion normally reserved for hummus ("toweling" in colloquial Hebrew), place the blackened clove on a piece of nice bread, and enjoy. Nowadays, far from the Middle East and partly housebroken by years of living in North America, I mostly get a nice enough Sicilian oil, but sorely miss the real thing.

Nutritionally, the pith is rich in a number of such known protective phytochemicals (substances that naturally occur in plants) as phenolics or flavonoids, as well as fiber, both soluble (notably pectin) and insoluble.[552,581,607,667] It's also rich in antioxidant and anti-inflammatory compounds[98,113,370] that improve lipid and carbohydrate metabolism, cognitive and mental health, osteoarthritis, and body weight, composition,

BRINGING PLANETARY EATING HOME: HOW TO SHOP, COOK, AND EAT 203

and fat percentage, and reduce the risk of cardiac events, stroke, and type II diabetes, Alzheimer's and other neurodegenerative diseases. These have been studied extensively and in many cases shown to confer significant protection,[376] including by "gold standard" placebo controlled, double blind randomized controlled trials.[250,436,533,658]

While surprising to some, these demonstrated health benefits are not entirely novel, but add a stamp of scientific rigor to interventions that have been widely used in traditional medicine for millennia.[348,544] As is true of most naturally occurring substances, citrus pith also likely contains currently unidentified compounds with unknown physiological influences. Such health promoting *micro*nutrients seem to me far more impactful for most modern overfed people's health, a far more worthy focus of our nutritional attention than our macronutrient balance (the fraction of total ingested calories that comes from fat, protein, or carbohydrates), an unrivaled, and unworthy which nonetheless star of the diet—health nexus. While some recommending bodies[198,577,599] have provided guidelines, to me they seem mostly like a needlessly dogmatic solution to an uninteresting, mostly solved problem, because our bodies can clearly handle a very wide range of these ratios.[303] While older age likely matters,[118,535] the required additions are modest, 5–10 g protein per day for an average-sized older adult.[149] Far more important than fruitlessly chasing the "ideal" macronutrient ratio are the origins, quality, and accompanying micronutrients of the macronutrients we ingest,[394,532] and what replaces the intake of any one macronutrient after any dietary modification (which is inevitable because increased intake of one macronutrient necessarily means ingesting less of another).

Returning to the self-defeating custom of discarding ever more food components, let's consider next removal of the paper thin skin of peanuts, pistachios, hazelnuts, or almonds, sometimes called "blanching." Pistachio skin is rich in such bioactive compounds as antioxidant phenolics, flavonoids, tannins, and fiber,[6,601] and almond skin contains 70–100% of the almonds' total phenols,[497] all known protective molecules most of us don't get enough of. Similarly, while the benefits of nut consumption are well established, consuming hazelnut skin appears to lower circulating blood levels of total fat, LDL cholesterol, triglycerides, and free fatty acids just as powerfully as the whole nut.[88] Or, dear to any Middle Eastern heart, tahini. Most commercial tahini is a pitiful cloudy liquid stripped bare of almost all nutritious solids, with greatly reduced fiber, protein, and phenol content and capacity to bind to and thus neutralize free oxygen radicals and LDL cholesterol. [232,233,538] Here's a straightforward shopping guide: if your candidate tahini has 1–2 g fiber per 200 kcals, avoid, but 5–6 g fiber per 200 kcals spell "actual tahini; buy."

Continuing our breakfast discussion, suppose you decide to make pancakes instead, thrusting wheat to the fore or, more specifically, the bizarre perverseness of wheat flour. One input, wheat kernels, two possible outputs, whole wheat or white flour. While both require milling, white flour also uses mechanized removal of the bran and germ, incurring energy, greenhouse gas, and financial costs. Yet the industrially degraded (white) flour is cheaper and more ubiquitous than its nutrient richer cousin. Why? The answer is as simple as it is twisted: with bran and germ removed, the remaining white flour is virtually fat free, eliminating flour's principal spoilage pathways, rancidity. For producers and marketers, this conveniently greatly extends the shelf life of the resulting white flour. For us the eaters, however, this removal of most that is nutritious in wheat and concentration of rapidly absorbed carbohydrates results in a food item, white flour, that is seven parts low quality food and three parts a vehicle for precision delivery of type II diabetes. [27,656]

All of the above examples and so many others in which parts of our food are mindlessly discarded are clear cases of the best interests of processors, distributors, or food retailers usurping our own, because in almost all instances, the post removal product is clearly inferior to the intact one in taste and nutritional value. Consider again total fiber, which is important both in its own right, and as a "canary in a coalmine" indicator of industrial food degradation. Based on epidemiological data, the US Institute of Medicine and Department of Agriculture recommend[577] daily total fiber intake of at least 14 g per ingested Mcal, or 27–34 g a day for most adults. Yet western diets tend to provide no more than half of this minimum, [317,546] compromising our health. [606,659] Modern agricultural mega-corporations, food processors, and supermarkets have thus cocreated a field-to-stomach food path that spares them various minor inconveniences, notably smaller, more frequent shipments, at the cost of aggressively promoting mostly preventable diet-related degenerative diseases. In the US, they have been abetted by the USDA, the obedient, meek, yet omnipresent frontperson of this band, but replicas of this defanged body in other nations sadly abound.

In keeping with the above "canary in a coalmine" status, the above dim view of food industrial processing is based on more than fiber only. In fact, content of each of the micronutrients the USDA[577] considers us deficient in—calcium, magnesium, potassium, choline, and vitamins A, C, D, and E—are thus diminished. The average US adult gets only 65–84% of those, regardless of supplementation status (figure 1 of Cowan et al.[132]), and even less among smokers and the food insecure (this is why it may be wise to supplement, especially if you are no longer young[392,665]). Because these removals undermine our health, it makes sense to favor, whenever possible, foods from which the least naturally occurring content was removed. (A complement rule, addressing concerns we have not addressed above, is: whenever possible, choose food with the least additions of sugar and salt.[426])

There are also arguments *against* eating the white citrus pith. The most pressing is toxic agrochemical residues, which—while in general less concentrated in the pith than on the peel itself[89]—may still be a concern. This raises another hot issue, the choice of organic vs conventional foods. One commonly invoked yet imperfectly established motivation for favoring organic fruits and vegetables is that they sometimes have lower residues of fewer industrial toxics than conventionally grown foods,[32,222,305,468] leading to markedly lower levels of such residues in urine samples of organic produce eaters.[135,360] Beyond statistics, this is entirely expected given that organic standards disallow several common agrochemicals, lending more weight to the above observations. But even if they all hold under required further scrutiny, it is still possible that conventional fruits or vegetables are perfectly safe because while their levels of toxic residues are higher than those of their organic counterparts, those elevated levels are still too low to pose health risks.[390] This speculation is indirectly supported by the health protective effects of fruits and vegetables regardless of organic status epidemiological studies reveal[190,624]; if you still reap the health benefits of fruits and vegetables even when they are conventional, conventional hazards are probably not too serious. Some evidence also suggests that this is indeed the case in the US (Appendix E of a report by the USDA[611]), but the issue is not settled,[282,512] partly because there are hundreds of agrochemicals, for many of which safe exposure limits are unknown.

My own synthesis of this information is to favor organic produce for items known to contain relatively high levels of toxic residue (such as strawberries, blueberries, bananas, hot and bell peppers, or various leafy greens[611]) when availability and budget permit (organic produce is often more expensive than its conventional counterparts). For dry raw cereals (oats, wheat) or legumes (soy, lentils), I don't bother with organic because it is almost universally unnecessary. This generalizes into a crude yet useful rule of thumb: if the product in question is cheap enough, it is unlikely to contain much toxic residue. For one thing, these crops are densely seeded, with no between-row spaces that can accommodate tractor wheels, excluding ground based spraying. Additionally, they tend to be semi-extensive, with low profit margins that disincentivize using costly agrochemicals. Oats offer a pertinent example. They have very little toxic residue,[611] and at the time of writing, the raw oat groats I use for breakfast and salads run about 65–70 cents a pound. However crude (and it *is* crude), this rule also applies to wheat, buckwheat, lentils, and chickpeas, among other crops.[611]

But you may wish to buck my rule for worthy reasons other than toxicity. An excellent example is financially encouraging operators who pursue exceptionally promising developments or are already actively playing positive roles, and whose products also happened to be organic. A particularly exciting example of this addresses the efforts to perennialize various cereals, legumes, and oilseeds (i.e., use traditional selection to compel them to regrow several successive seasons) at, for example, Washington State University or The Land Institute in Kansas. Since perennials are not annually reseeded or replanted, perennialized crops are poised to prevent much plowing and herbicide applications, thus promoting less toxic landscapes, improved soil health, carbon sequestration, and more efficient water utilization. (But in some cases, because perennials limit mechanical cultivation, they in some cases may actually enhance toxic sparing.)

However, the most potentially damning environmental argument against organic is reduced yield,[473] estimated to be 10–20%[141] to, occasionally, as high as 30–40%.[318,323] Such yield gaps may not be universal,[399] depending instead on operators' experience and the suitability to specific environments of cultivars (specific genotypes of a given crop) for organic production.[66] But suppose organic yield gaps—which raise demand for cropland and likely other resources per kilogram of product—are real and enduring, and suppose they are as high as 40% (or, equivalently, that the cropland needs of organic cultivation are 65–70% higher than those of conventional cultivation, see Accompanying PDF). Then, people switching to organics while maintaining their pre-conversion diets would require more resources, which is a strong argument against organic. Luckily, this heightened resource need is easily overcome by reducing meat intake (see Eshel et al.[183] and Accompanying PDF). For example, reducing by one third this intake from its current levels (reducing meat's portion in our total protein intake from the current portion, 43%, to 29%) while increasing plant protein consumption so as to fully replenish the lost meat protein would fully offset the elevated land needs due to the presumed organic yield gap, yielding a diet containing the same amount of protein we currently eat, but with much more known protective nutrients (see ref.[183] and the Accompanying PDF for details).

A modest replacement of meat with non-meat alternatives can thus completely undo the added cropland needs associated with favoring organic over conventional produce. This strongly counters the potentially most persuasive argument against organic, its supposed yield gap. More generally, this result calls into question a key tenet of sustainable intensification, the broad idea that food production must rise to meet rising demand, and do so mostly using existing cropland.[99,280] Recognized or not, the sustainable intensification ethos implies that some environmental costs—notably environmental toxicity due to agrochemical use—are inevitable in a food system yoked to the expanding production wagon, and must be managed. Since organic production, as we have seen, is associated with lower usage of fewer agrochemicals, this ethos also implies a limited role to organic production as long as its yield gap persists. While sustainable intensification is compelling, the above result highlights that it is so *only* when it also aggressively promotes dietary shifts. Unfortunately, whereas scientific promoters of sustainable intensification are well aware of and explicitly emphasize this duality, practical interpretations of the idea often neglect dietary shifts.[564] Absent a firm commitment to such shifts, sustainable intensification is hallowed of environmental meaning.

Having explored some hotly debated topics—discarding perfectly good parts of our foods, micronutrient sufficiency, organic vs conventional, sustainable intensification—it's time to begin making our finocchio salad.

First, we'll cook lentils and quinoa (or farro; they are nicely interchangeable, and in fresh salads I usually mix black, green, and red lentils, for aesthetics). As already discussed, organic is unimportant for either of these cereals and legumes; there is no difference that I can discern between what you get in artfully wasteful packages and their half-price alternatives.

Next comes choosing the cooking method. This is very important, because for pulses, the energy used in home cooking dominates the total field-to-plate environmental costs, accounting for 75–90% of the emissions, water consumption, and water pollution (figure 3 of Bandekar et al.[33]). But note that while cooking matters greatly here in terms of relative contributions to total environmental costs, our salad is going to be super carbon efficient. My estimate is we'll only use up around 0.8–0.9 kg CO_{2eq} per serving containing 17–19 g of protein. By comparison, producing that protein mass from beef would have required emissions of 4–10 kg CO_{2eq}[179,180] or more,[278,475] and this is only to the farm gate, excluding emissions due to packaging, distribution, retail, and cooking).

So, how do we use the least energy when cooking our lentils, farro or quinoa? Disappointingly, few published comparisons exist, and I have found none I trust. But the May 2, 2012 "Burning Desire for Efficiency" post on dothemath.ucsd.edu is just what we need, a carefully and meticulously researched gem by Tom Murphy, a physics professor at UC San Diego. Because these are really simple measurements to make, and Tom is a highly reputed practical experimentalist whose informal writing stands beautifully the test of time, I trust Tom's numbers (as does, apparently, *The New York Times*[496]). Tom's measurements show that with its 15–30% efficiency, a gas range is deplorable. This means that of every 100 joules liberated by the flame, only 15–30 end up heating the water, with the rest needlessly heating the kitchen. Terrible though this is, it ignores all that happened to this gas before it got into your tank, which adds 20–25%,[602] dropping the efficiency to 13–26%. Gas is definitely out.

What's in is electric; traditional resistance-based range, microwave, or induction range. Tom's blog shows a modest and uncertain edge for a resistance-based electric kettle over a microwave oven, about 50% and 40% efficiency, respectively. But it also shows that if you ditch the auto toggle and watch the kettle (exactly as the sagacious saying tells you *not* to do) and turn it off as soon as the water first boils, your kettle will achieve 80% efficiency. My own estimate in our kitchen led to a 93% estimated efficiency, but I went by the kettle's nominal wattage, making Tom's measurements more robust. So, here's the plan: since we will be cooking 2 cups lentils and 1 cup quinoa, we're going to put 7 cups of water in the kettle and turn it on. Next, we will put each lentil type, and the quinoa, in empty soup bowls. The water need not actually boil; lentils don't need it, quinoa certainly doesn't, and because the specific heat of water (c_p, a measure of the energy needed to raise one kilogram of a given substance by $1°C$) rises as temperature nears boiling, the unrequired $98°–100°C$ part of the journey is more energy intensive than the $90°–92°C$ one. Once the kettle makes that low groan that announces imminent boiling, we will—following Tom's helpful experiments—turn it off immediately, pour the water into the bowls (cover plus some), and put the bowls in the microwave oven. Since the bowls and their content likely dropped to $80°–85°C$, let's zap them, about ninety seconds (in my 1,100 W microwave), only until gentle surface undulation first appears. And that's it for now; I just did exactly the above, and came right back to writing. Some forty-five minutes later, the quinoa is ready. I drink the excess water if there is any (some of the protein dissolves in the cooking water), but you don't have to. The lentils would probably require another, but shorter, go of the microwave. I do all of this right after cleaning up the kitchen after breakfast, and let the lentils sit in their hot, then warm water, pondering. With the exception of solar ovens or residual heat from some other appliance that produces byproduct heat, this is as energy efficient as I can think of in most kitchens (but see the Accompanying PDF for a discussion on winter cooking on wood burning stoves, with a special nod to the great George Monbiot, [404] with whom I hereby respectfully disagree).

After the lentils and quinoa have sat in the microwave in slowly cooling water for several hours, I take out the quinoa, use the microwave to reheat the lentil bowls to 85°–90°C, and let them sit for ten to fifteen additional minutes. (This is how I cook and do all house chores: break everything into short steps and carry them out during short breaks from writing or calculating. But I am mindful that this is not generalizable beyond the privileged professor set.) At the next check-in, I layer the bottom of a colander with fresh spinach leaves, and pour the hot lentils with the water on top of the spinach, fully covering it. I capture the draining liquid, cloudy and viscous due to its the high protein content, for later use. One is to replace eggs in muffins or waffles (this is partly why I add no salt to the cooking water). Another is a nice, simple winter breakfast: this liquid, a bit of miso paste, cubed extra firm tofu, chopped scallion, green cabbage bits, thinly sliced ginger root, a drop of toasted sesame oil, whole red chili peppers, and nori if I have any. Kimchi makes for great if optional flotsam in this brew, but when present, cut the miso markedly or it'll get too salty.

If you wonder about such peculiar eating habits, here is the abstract. In the early 1980s, I worked for a few years on oil rigs in the Gulf of Mexico. This is where my eating habits came of age. For every ten men (and it *was* only men then), there were eleven languages spoken, and twelve different cuisines, with the only common language being the hand gestures universal in commercial diving. Foodwise, it was revolutionary after growing up on mostly finely diced tomatoes, cucumbers and onion, with oil, olives, eggs, cheese. Here were people blending ingredients I have never heard of with choice parts of the oceanic foodweb they caught in the clear blue Gulf water or scraped off the rig's steel "legs." Proud of their culinary traditions, they were mostly eager to share their food, which I sampled voraciously, helped by pantomime instructions on the proper ways to eat it. There were also regular visits to nearby Mexican towns. I had my bike with me, and I rode the back roads between Veracruz and the Yucatan peninsula often and at length. There—along huge swamps, home to gray pelican flocks vast enough to darken the midday sky when they took flight as one—I encountered sprawling fresh markets, offering yet more new foods and amazing ways to eat them. All of those meals, on sea or land, blended into the peculiar mix that is my current eating habits.

Back to the spinach wilting under the slowly cooling legumes and cereals. In the next and final work break, I place in a large wooden salad bowl

BRINGING PLANETARY EATING HOME: HOW TO SHOP, COOK, AND EAT

a medium thinly sliced purple onion, 4–5 generous tablespoons of raw tahini, juice of 2 large lemons, 2–3 tablespoons each of balsamic and apple cider vinegar, olive oil (to taste; lots), black pepper, chili peppers, a pinch of oregano and another of thyme, and some miso paste, mix, and let sit. After washing and drying the large finocchio head, I trim the root side end very close to the bottom, and work my way up deep into the green. I discard almost nothing, unless the top stocks are aged and woody, with visible vascularity. The thin dill-like hairs are great, finely chopped and added to the bowl. Mix in and let sit together for about an hour. Note that the ratio of tahini to lemon juice and vinegar is too low for oil in water emulsion, the normal state of tahini, to form. The roasted sesame butter therefore persists in its raw state, playing a very different physical and culinary role than normal tahini.

After another work bout, it is time to get serious. We add the lentils, quinoa, and spinach to the bowl in which the onion, fennel, and spices were sitting, and warm a bit of olive oil in a pan, for sauteing thinly sliced onion until translucent. While this is happening, we thinly slice 8–10 black olives in the bowl. Back to the pan, when the onion is almost ready, add thin slices of garlic, 1–3 cloves to taste; 15 seconds, stir, another 15, and the whole thing goes into the bowl. With the pan back over medium heat, we add in brown hazelnuts and stir constantly for 1–2 minutes. Add a fistful of pine nuts next, and keep stirring for another minute or so, but watch the pine nuts like a hawk, they turn from underdone to burnt in seconds. Once more, add to the bowl and turn off the range. Add a whole bunch each of chopped fresh mint and dark green flat parsley (avoid the regrettable curly type).

It is time for final touches. For spices, I go easy. Definitely no additional salt; the miso and olives are plenty salty for most. I almost always add pumpkin seeds in the shell. They are delicious, and add crunch, fiber, and protein (but even without them, this salad is very fiber and protein rich). Their presence makes people chew much more methodically, and thus eat more slowly, always a plus. I also love to add (unsalted!) pistachios, but if you do, note that their crunch rapidly deteriorates in the wet salad environment. Pecans (not a fave here, but hey...), sunflower seeds, black and/or white sesame seeds, or hemp seeds, all make nice optional additions.

Some like the fruit route. To take it, add some cubed orange, grapefruit, or apple. As discussed earlier, it is best to keep the pith on, and apple cores (minus the seeds) in. In winter, you can add very finely chopped dried tart cherries, currants, or some dried cranberries. In summer, diced plum, apricot, or peach are all nice. For dairy eaters, cubed feta works well. If you are feeling ambitious, and can get any, fry cubed haloumi in olive oil. When all sides are light brown, add to the salad.

My guests almost universally like my many different permutations of the above. I hope you and yours do too. Yet, my purpose here is not to show off my decidedly modest kitchen skills but to highlight practical manifestations of the main related yet separate threads we have covered. In light of this objective, *what does this salad teach us?*

The current food system fails us in multiple deep, pervasive ways. It promotes incidence of diet-related degenerative diseases well beyond natural rates. By producing 20–30% of all greenhouse gas emissions, it damages many natural systems our well-being depends on, further undermining global safety and health. By dominating coastal water pollution, it pits upstream industrial farmers against downstream fishers and pronounces the latter losers, degrading the economic livelihood of coastal communities and further undermining their health. By also dominating inland freshwater pollution, it inflicts similar harms on communities along most rivers and lakes. By dominating consumptive fresh water use in many arid basins in the western US and worldwide, the food system deprives local residents and wildlife of this highly critical and severely limited resource, and replumbs nature by re-engineering terrestrial eco-hydrology and hydrometeorology. In almost all cases, these re-engineered systems are pale, semi-functional imitations of the systems they replace.

Some of this multifaceted damage is the inevitable cost of feeding ourselves. But much of it is not, stemming instead from the foundational principles of agribusiness and the agricultural policies that enable them: keep the price of raw agricultural products well below their societal costs and maximize profits by maximizing "value addition" via processing. Unfortunately, keeping agricultural products very cheap is achieved through the pervasive use of agricultural practices—overusing tilling, synthetic fertilizer, pesticides, herbicides, and other agrochemicals—that are lifesavers when used judiciously but the principal conduits of the above destruction today. And the inevitable inextricable companion of the processing that is the beating heart of agribusiness is severe degradation of nutritional value, turning invaluable sustenance into disease vectors.

Modern societies clearly need agribusiness, without which diets will be meager, erratic, and inaccessible. But, as the spectacular demise of the Sackler family attests, societal acceptability requires businesses to balance profit maximization with protecting our health. Yet, in today's agribusinesses, profit maximization is unopposed, relegating nutrition and environmental conservation to vacuous platitudes in corporate brochures. The social contract between governments, agribusiness, and—most importantly—we the eaters must be renegotiated, this time with the natural constraints within which these systems function, our bodies and the physical environment, taking center stage. Some of this renegotiating will inevitably take place in conference rooms and legislative halls, but even if these deliberations bear fruit, their impacts are a distant future. Immediate impact—safeguarding our personal health and the environment in which we dwell *now*—requires distinct additional strategies.

One such strategy emphasizes personal choice, opting out of continuing to subserviently play the role assigned to us by the corporations that lord over this lopsided system. With this strategy in mind, our salad transforms from a humble meal to a blueprint for a rebellion. Using our finocchio salad and its countless culinary kin to make low profit, barely processed agricultural outputs the foundation of our diet is one fine strategy for restoring balance to the social contract mentioned above between governments, agribusiness, and eaters. It uses capitalist market forces to compel governments to rethink keeping agricultural products far cheaper than their actual societal costs, and agribusiness to drastically demote hyper processing in their foundational business model. Not merely delightful, our salad is a political statement that, like lunch counter sit-ins, sends a clear message that we hereby resign our assigned role in this farce, because it sickens us, in more ways than one.

To be adapted widely enough for success, this approach must possess the following key attributes.

Most importantly, uprising food must be delicious. While I think the recipe above is, I am clearly a mediocre amateur. Turning raw, minimally processed commodities—lentils, chickpeas, cereals, beans—into delicious, widely availed real food without obscene price tags or nutritional degradation is where real cooks can shine. We do not need their environmental musings; excepting Mark Bittman, those are often naive and uninformed. Fortunately, they also almost universally come with just the laughable price tags—www.exploretock.com estimates \$350–\$400 per person per Stone Barns meal—to guarantee zero impact, beautifully produced treaties to the contrary notwithstanding.[35] But we urgently need good, honest cooks who develop and perfect such recipes as our finocchio salad to become highly popular, which experience[59] shows can be done, yet which is currently sporadic at best. Meals that meet these criteria have been delicious, ubiquitous, and cheap in any Middle Eastern town continuously and unfailingly since well before Jesus was born there. The same holds for different permutations in many cultures. Are the people of the US, France, or Germany so fundamentally less discerning than the denizens of these bygone towns that this is impossible there?!

Heeding the affordability pillar of the proposed rebellion, our salad is very cheap, \$2–3 per serving in summer 2023, and can be made cheaper still by bulk buying and skimping on some more expensive optional elements. For example, in New York, Berlin, or Paris, falafel and hummus are everywhere, not only because they taste so good but also because of the superior nutrition and affordability they offer.

Which brings us to the final pillar of the proposed rebellion, health. This salad and similar meals will, statistically speaking, extend your life, and reduce disease during those longer years. All the health-related notions invoked to promote the necessity of animal products are houses of empirical cards. Animal protein is indeed high quality and amino acid diverse, as is our salad; with the exception of vitamin B_{12}, animal products offer no nutrient plants cannot supply at far lower environmental costs; beef is indeed a rich source of iron, of which most developed nation denizens other than menstruating-age women need less.

In short, if you are, like countless others, disillusioned with our food system, here is a nice, delicious, practical alternative to the output it favors. Not only is it far superior—nutritionally, environmentally and, most importantly, flavor-wise—but it also puts agribusinesses on notice that remaining in the business of promoting disease and environmental degradation requires finding a new population to undermine, because the denizens of *this* Earth choose life over agribusiness bottom line, health and working bodies over elective degenerative disease, and actual planetary betterment over the vacuous environmental platitudes typically promulgated by these businesses.

Bibliography

[1] S. Abrahamyan, Z. Ahmed, H. Albataineh, et al. Measurement of the neutron radius of ^{208}Pb through parity violation in electron scattering. *Physical Review Letters*, 108(112502), 2012.

[2] M. Adil, S. Zhang, J. Wang, et al. Effects of fallow management practices on soil water, crop yield and water use efficiency in winter wheat monoculture system: A meta-analysis. *Frontiers in Plant Science*, 13, 2022.

[3] S. B. Aher, B. L. Lakaria, A. B. Singh, et al. Nutritional quality of soybean and wheat under organic, biodynamic and conventional agriculture in semi-arid tropical conditions of Central India. *Indian Journal of Agricultural Biochemistry*, 31:128–136, 2018.

[4] L. C. Aiello. Brains and guts in human evolution: The expensive tissue hypothesis. *Brazilian Journal of Genetics*, 20, 1997.

[5] G. E. Aiken. Grassfed beef from a global perspective. Technical report, United States Department of Agriculture, Agricultural Research Service, Forage-Animal Production Research Unit, Lexington, KY, 2012.

[6] C. Alasalvar and B. W. Bolling. Review of nut phytochemicals, fat-soluble bioactives, antioxidant components and health effects. *British Journal of Nutrition*, 113(S2):S68–S78, 2015.

[7] J. Aleon, D. Levy, A. Aleon-Toppani, et al. Determination of the initial hydrogen isotopic composition of the solar system. *Nature Astronomy*, 6:458–463, 2022.

[8] M. J. Alessi, D. A. Herrera, C. P. Evans, et al. Soil moisture conditions determine land-atmosphere coupling and drought risk in the northeastern

United States. *Journal of Geophysical Research: Atmospheres*, 127(6):e2021JD0347 40, 2022.

[9] C. M. O'D. Alexander, R. Bowden, M. L. Fogel, et al. The provenances of asteroids, and their contributions to the volatile inventories of the terrestrial planets. *Science*, 337(6095):721–723, 2012.

[10] M. R. Allen, K. P. Shine, J. S. Fuglestvedt, et al. A solution to the misrepresentations of CO_2-equivalent emissions of short-lived climate pollutants under ambitious mitigation. *NPJ Climate and Atmospheric Science*, 1(16), 2018.

[11] R. E. Alter, H. C. Douglas, J. M. Winter, et al. Twentieth century regional climate change during the summer in the central United States attributed to agricultural intensification. *Geophysical Research Letters*, 45(3):1586–1594, 2018.

[12] M. Amann, G. Kiesewetter, W. Schöpp, et al. Reducing global air pollution: The scope for further policy interventions. *Philosophical Transactions of the Royal Society A: Mathematical, Physical and Engineering Sciences*, 378(2183):20190331, 2020.

[13] T. Amano, N. Gerrett, Y. Inoue, et al. Determination of the maximum rate of eccrine sweat glands' ion reabsorption using the galvanic skin conductance to local sweat rate relationship. *European Journal of Applied Physiology*, 116(2):281–290, 2016.

[14] R. Amundson and L. Biardeau. Soil carbon sequestration is an elusive climate mitigation tool. *Proceedings of the National Academy of Sciences*, 115(46):11652–11656, 2018.

[15] S. I. Apfelbaum, R. Thompson, F. Wang, et al. Vegetation, water infiltration, and soil carbon response to adaptive multi-paddock and conventional grazing in southeastern USA ranches. *Journal of Environmental Management*, 308(114576), 2022.

[16] D. Archer. Fate of fossil fuel CO_2 in geologic time. *Journal of Geophysical Research*, 110(C09S05), 2005.

[17] D. Archer and V. Brovkin. The millennial atmospheric lifetime of anthropogenic CO_2. *Climatic Change*, 90:283–297, 2008.

[18] D. Archer, M. Eby, V. Brovkin, et al. Atmospheric lifetime of fossil fuel carbon dioxide. *Annual Review of Earth and Planetary Sciences*, 37:117–134, 2009.

[19] A. Arcones and F. K. Thielemann. Origin of the elements. *Astronomy and Astrophysics Review*, 31(1), 2023.

[20] B. Arends. "Totally bizarre!"—nutritionists see red over study downplaying the serious health risks of red meat. 1, 2020.

BIBLIOGRAPHY

[21] D. R. Aryal. Grazing intensity in grassland ecosystems: Implications for carbon storage and functional properties. *CABI Reviews*, 17(032):1–12, 2022.

[22] P. Ashwin and A. S. von der Heydt. Extreme sensitivity and climate tipping points. *Journal of Statistical Physics*, 179:1531–1552, 2020.

[23] S. A. Asmarasari, N. Azizah, S. Sutikno, et al. A review of dairy cattle heat stress mitigation in Indonesia. *Veterinary World*, 16:1098–1108, 2023.

[24] D. J. Augustine, J. D. Derner, L. M. Porensky, et al. Adaptive, multi-paddock, rotational grazing management: An experimental, ranch-scale assessment of effects on multiple ecosystem services. In *Proceedings of the XXIV International Grassland Congress / XI International Rangeland Congress (Sustainable Use of Grassland and Rangeland Resources for Improved Livelihoods)*, pp. 1–4, Nairobi, Kenya, 10 2021. Kenya Agricultural and Livestock Research Organization.

[25] D. J. Augustine, J. D. Derner, M. E. Fernandez-Gimenez, et al. Adaptive, multi-paddock rotational grazing management: A ranch-scale assessment of effects on vegetation and livestock performance in semiarid rangeland. *Rangeland Ecology & Management*, 73:796–810, 2020.

[26] D. E. Ausband and L. D. Mech. The challenges of success: Future wolf conservation and management in the United States. *BioScience*, 73:587–591, 2023.

[27] P. S. Baenziger, K. Frels, S. Greenspan, et al. A stealth health approach to dietary fibre. *Nature Food*, 4:5–6, 2023.

[28] Y. Bai and M. F. Cotrufo. Grassland soil carbon sequestration: Current understanding, challenges, and solutions. *Science*, 377, 2022.

[29] S. K. Bajgain, A. W. Ashley, M. Mookherjee, et al. Insights into magma ocean dynamics from the transport properties of basaltic melt. *Nature Communications*, 13(7590), 2022.

[30] S. K. Bajgain, M. Mookherjee, R. Dasgupta, et al. Nitrogen content in the earth's outer core. *Geophysical Research Letters*, 46(1):89–98, 2019.

[31] D. R. Bakaloudi, A. Halloran, H. L. Rippin, et al. Intake and adequacy of the vegan diet. A systematic review of the evidence. *Clinical Nutrition*, 40(5):3503–3521, 2021.

[32] B. P. Baker, C. M. Benbrook, E. Groth III, et al. Pesticide residues in conventional, integrated pest management (ipm)-grown and organic foods: Insights from three US data sets. *Food Additives & Contaminants*, 19(5):427–446, 2002.

[33] P. A. Bandekar, B. Putman, G. Thoma, et al. Cradle-to-grave life cycle assessment of production and consumption of pulses in the United States. *Journal of Environmental Management*, 302:114062, 2022.

[34] E. Bangs and J. A. Shivik. Managing wolf conflict with livestock in the northwestern United States, 7, 2001.

[35] D. Barber. *The Third Plate: Field Notes on the Future of Food*. Penguin, New York, 2015.

[36] J. L. Barnard. The bison and the cow: Food, empire, extinction. *American Quarterly*, 72, 2020.

[37] K. A. Barnes, M. L. Anderson, J. R. Stofan, et al. Normative data for sweating rate, sweat sodium concentration, and sweat sodium loss in athletes: An update and analysis by sport. *Journal of Sports Sciences*, 37:2356–2366, 2019.

[38] J. L. Batchelor, W. J. Ripple, T. M. Wilson, et al. Restoration of riparian areas following the removal of cattle in the Northwestern Great Basin. *Environmental Management*, 55:930–942, 2015.

[39] N. H. Batjes. Technologically achievable soil organic carbon sequestration in world croplands and grasslands. *Land Degradation & Development*, 30:25–32, 2019.

[40] P. Bauman and A. Williams. Grass-fed beef: Market share of grass-fed beef. Technical report, South Dakota State Unoversity Extension, 6 2021.

[41] K. A. Beauchemin and S. M. McGinn. Methane emissions from feedlot cattle fed barley or corn diets. *Journal of Animal Science*, 83:653–661, 3 2005.

[42] N. Becerra-Tomás, N. Babio, M. Á. Martínez-González, et al. Replacing red meat and processed red meat for white meat, fish, legumes or eggs is associated with lower risk of incidence of metabolic syndrome. *Clinical Nutrition*, 35(6):1442–1449, 2016.

[43] C. A. Becker, R. J. Collier, and A. E. Stone. Invited review: Physiological and behavioral effects of heat stress in dairy cows. *Journal of Dairy Science*, 103:6751–6770, 2020.

[44] G. Bedenham, A. Kirk, U. Luhano, et al. The importance of biodiversity risks: Link to zoonotic diseases. *British Actuarial Journal*, 27(E10), 2022.

[45] D. Beillouin, J. Demenois, R. Cardinael, et al. A global database of land management, land-use change and climate change effects on soil organic carbon. *Scientific Data*, 9(228), 2022.

[46] P. Benaud, K. Anderson, M. Evans, et al. National-scale geodata describe widespread accelerated soil erosion. *Geoderma*, 371:114378, 2020.

[47] E. J Benjamin, P. Muntner, A. Alonso, et al. Heart disease and stroke statistics—2019 update: A report from the American Heart Association. *Circulation*, 139(10):e56–e528, 2019.

[48] M. Bennett-Smith. From model t to prius: 13 big moments in fuel efficiency history. *Christian Science Monitor*, 3, 2012.

BIBLIOGRAPHY

[49] K. Benoit. Regenerative ranching: Analyzing the impact of specific variation in rotational grazing practices on pasture health and biodiversity. Master's thesis, University of Michigan, 2024.

[50] K. Berding and J. F. Cryan. Microbiota-targeted interventions for mental health. *Current Opinion in Psychiatry*, 35:3–9, 2022.

[51] A. Berman. Effects of body surface area estimates on predicted energy requirements and heat stress. *Journal of Dairy Science*, 86(11):3605–3610, 2003.

[52] A. M. Bernstein, Q. Sun, F. B. Hu, et al. Major dietary protein sources and the risk of coronary heart disease in women. *Circulation*, 122(9):876–883, 2010.

[53] W. Berry. *The Unsettling of America: Culture & Agriculture*. Counterpoint Press, Berkeley, CA, 2004.

[54] R. L. Beschta and W. J. Ripple. Riparian vegetation recovery in yellowstone: The first two decades after wolf reintroduction. *Biological Conservation*, 198:93–103, 2016.

[55] Z. Bian, H. Tian, Q. Yang, et al. Production and application of manure nitrogen and phosphorus in the United States since 1860. *Earth System Science Data*, 13:515–527, 2021.

[56] K. J. Biddinger, C. A. Emdin, M. E. Haas, et al. Association of habitual alcohol intake with risk of cardiovascular disease. *JAMA Network Open*, 5(3):e223849–e223849, 03 2022.

[57] D. Bigelow and A. Borchers. Major uses of land in the United States, 2012. EIB-178, 69 pp; Online at https://www.ers.usda.gov/publications/pub-details/?pubid=84879; accessed 14 June 2021.

[58] D. Biswas and N. S. Azad Thakur. Seasonal incidence of major insect pests of apple in mid hills of Meghalaya. *Indian Journal of Entomology*, 83:459–463, 2021.

[59] M. Bittman. *How To Cook Everything Vegetarian: A Plant-Based Vegetarian Cookbook*. Harvest, Eugene, OR, 97404-0322, 2017.

[60] M. Bittman. *Animal, Vegetable, Junk: A History of Food, from Sustainable to Suicidal*. Mariner Books, Boston, 2021.

[61] M. Bittman and D. Katz. *How To Eat: All Your Food and D. Questions Answered: A Food Science Nutrition Weight Loss*. Houghton Mifflin, Boston, 2020.

[62] A. R. Blaustein, J. M. Kiesecker, D. P. Chivers, et al. Ambient UV-B radiation causes deformities in amphibian embryos. *Proceedings of the National Academy of Sciences*, 94(25):13735–13737, 1997.

[63] D. Blaustein-Rejto, N. Soltis, and L. Blomqvist. Carbon opportunity cost increases carbon footprint advantage of grain-finished beef. *PLOS ONE*, 18:1–14, 2023.

[64] BLM. Gila box riparian national conservation area.

[65] L. Bloom. Letter. *New York Times*, 11, 2009. Letter to the Editor; author is of Easton, CT.

[66] E. Bodur, D. Kilic, and O. Caliskan. Effects of organic and conventional production systems on plant vigor, fruit yield and fruit quality attributes of bananas cultivated in the Mediterranean region of Turkey. *Erwerbs-Obstbau*, 65:143–152, 2023.

[67] E. W. Bork, T. F. Dobert, J. S. J. Grenke, et al. Comparative pasture management on Canadian cattle ranches with and without adaptive multipaddock grazing. *Rangeland Ecology and Management*, 78:5–14, 2021.

[68] P. Borrelli, D. A. Robinson, L. R. Fleischer, et al. An assessment of the global impact of 21st century land use change on soil erosion. *Nature Communications*, 8(2013), 2017.

[69] S. Bova, Y. Rosenthal, Z. Liu, et al. Seasonal origin of the thermal maxima at the holocene and the last interglacial. *Nature*, 589:548–553, 2021.

[70] A. J. Boyce, H. Shamon, and W. J. McShea. Bison reintroduction to mixed-grass prairie is associated with increases in bird diversity and cervid occupancy in riparian areas. *Frontiers in Ecology and Evolution*, 10, 2022.

[71] D. D. Briske, A. J. Ash, J. D. Derner, et al. Commentary: A critical assessment of the policy endorsement for holistic management. *Agricultural Systems*, 125:50–53, 2014.

[72] D. D. Briske, B. T. Bestelmeyer, and J. R. Brown. Savory's unsubstantiated claims should not be confused with multipaddock grazing. *Rangelands*, 36:39–42, 2014.

[73] D. D. Briske, B. T. Bestelmeyer, J. R. Brown, et al. The Savory method can not green deserts or reverse climate change. *Rangelands*, 35:72–74, 2013.

[74] L. F. Brito, N. Bedere, F. Douhard, et al. Review: Genetic selection of high-yielding dairy cattle toward sustainable farming systems in a rapidly changing world. *Animal*, 15(100292), 2021.

[75] C. Brock, U. Geier, R. Greiner, et al. Research in biodynamic food and farming—A review. *Open Agriculture*, 4:743–757, 2019.

[76] I. A. Brouwer, A. J. Wanders, and M. B. Katan, et al. Trans fatty acids and cardiovascular health: Research completed? *European Journal of Clinical Nutrition*, 67:541–547, 2013.

[77] V. Brovkin, E. Brook, J. W. Williams, et al. Past abrupt changes, tipping points and cascading impacts in the earth system. *Nature Geoscience*, 14:550–558, 2021.

[78] V. Brovkin, T. Brucher, T. Kleinen, et al. Comparative carbon cycle dynamics of the present and last interglacial. *Quaternary Science Reviews*, 137:15–32, 2016.

BIBLIOGRAPHY

[79] C. Brown, R. Prestele, and M. Rounsevell, et al. An assessment of future rewilding potential in the United Kingdom. *Conservation Biology*, Online pre-publication(e14276), 2024.

[80] E. Bruenig. Why do evangelicals like James Inhofe believe that only god can cause climate change? *New Republic*, 1, 2015.

[81] B. A. Buffett, E. J. Garnero, and R. Jeanloz. Sediments at the top of earth's core. *Science*, 290:1338–1342, 2000.

[82] C. Bugbee and R. Silver. Sixty-day notice of intent to sue pursuant to the endangered species act violations in the gila box riparian national conservation area in Arizona, 7, 2021.

[83] E. Bullard, J. E. Everett, A. G. Smith, et al. The fit of the continents around the Atlantic. *Philosophical Transactions of the Royal Society of London. Series A, Mathematical and Physical Sciences*, 258(1088):41–51, 1965.

[84] G. Buono, V. Massara Paletto, and D. Celdran. Forage availability dynamics of a Patagonian steppe under different grazing use intensities by sheep. *Revista Argentina de Produccion Animal*, 31:135–143, 2011.

[85] E. Margaret Burbidge, G. R. Burbidge, W. A. Fowler, et al. Synthesis of the elements in stars. *Reviews of Modern Physics*, 29:547–650, 1957.

[86] A. Burke. Hunting in the middle palaeolithic. *International Journal of Osteoarchaeology*, 10:281–285, 2000.

[87] C. Dermot and S. Elavarthi. Rangelands as carbon sinks to mitigate climate change: A review. *Journal of Earth Science & Climatic Change*, 5(1000221), 2014.

[88] A. Caimari, F. Puiggròs, M. Suárez, et al. The intake of a hazelnut skin extract improves the plasma lipid profile and reduces the lithocholic/deoxycholic bile acid faecal ratio, a risk factor for colon cancer, in hamsters fed a high-fat diet. *Food Chemistry*, 167:138–144, 2015.

[89] E. Calvaruso, G. Cammilleri, A. Pulvirenti, et al. Residues of 165 pesticides in citrus fruits using lc-ms/ms: A study of the pesticides distribution from the peel to the pulp. *Natural Product Research*, 34:34–38, 2020.

[90] C. Cambeses-Franco, S. González-García, G. Feijoo, et al. Is the paleo diet safe for health and the environment? *Science of the Total Environment*, 781:146717, 2021.

[91] P. Canning. A revised and expanded food dollar series: A better understanding of our food costs. Technical Report Economic Research Report Number 114, U. S. Department of Agriculture, Economic Research Service, February 2011.

[92] J. L. Capper. Is the grass always greener? Comparing the environmental impact of conventional, natural and grass-fed beef production systems. *Animals*, 2:127–143, 2012.

[93] J. L. Capper. The environmental impact of beef production in the United States: 1977 compared with 2007. *Journal of Animal Science*, 89:4249–4261, 2011.

[94] J. L. Capper. Is the grass always greener? Comparing the environmental impact of conventional, natural and grass-fed beef production systems. *Animals*, 2:127–143, 2012.

[95] C. J. Carey, J. Weverka, R. DiGaudio, et al. Exploring variability in rangeland soil organic carbon stocks across California (USA) using a voluntary monitoring network. *Geoderma Regional*, 22:e00304, 2020.

[96] J. Carey. Core concept: Rewilding. *National Academy of Sciences*, 113(4):806–808, 2016.

[97] J. Carter, A. Jones, M. O'Brien, et al. Holistic management: Misinformation on the science of grazed ecosystems. *International Journal of Biodiversity*, 2014(163431):10, 2014.

[98] R. Casquete, S. M. Castro, M. C. Villalobos, et al. High pressure extraction of phenolic compounds from citrus peels†. *High Pressure Research*, 34(4):447–451, 2014.

[99] K. G. Cassman and P. Grassini. A global perspective on sustainable intensification research. *Nature Sustainability*, 3:262–268, 2020.

[100] D. S. M. Chan, R. Lau, D. Aune, et al. Red and processed meat and colorectal cancer incidence: Meta-analysis of prospective studies. *PLoS One*, 6(e20456), 2011.

[101] J. Chang, P. Ciais, T. Gasser, et al. Climate warming from managed grasslands cancels the cooling effect of carbon sinks in sparsely grazed and natural grasslands. *Nature Communications*, 12(118), 2021.

[102] V. Chaplot, P. Dlamini, and P. Chivenge. Potential of grassland rehabilitation through high density-short duration grazing to sequester atmospheric carbon. *Geoderma*, 271:10–17, 2016.

[103] J. Charney, W. J. Quirk, S.–h. Chow, et al. A comparative study of the effects of albedo change on drought in semi–arid regions. *Journal of the Atmsopheric Sceinces*, 34:1366–1385, 1977.

[104] S. Charnley, H. Gosnell, R. Davee, et al. Ranchers and beavers: Understanding the human dimensions of beaver-related stream restoration on western rangelands. *Rangeland Ecology and Management*, 73:712–723, 2020.

[105] A. Chaudhary, S. Pfister, and S. Hellweg. Spatially explicit analysis of biodiversity loss due to global agriculture, pasture and forest land use from a producer and consumer perspective. *Environmental Science & Technology*, 50(7):3928–3936, 2016.

BIBLIOGRAPHY   225

[106] A. S. Chavez and E. M. Gese. Food habits of wolves in relation to livestock depredations in northwestern Minnesota. *The American Midland Naturalist*, 154(1):253–263, 2005.

[107] B. Chen, M. Y. Han, K. Peng, et al. Global land-water nexus: Agricultural land and freshwater use embodied in worldwide supply chains. *Science of the Total Environment*, 613-614:931–943, 2018.

[108] C. Chen, I. Noble, J. Hellmann, et al. University of Notre Dame global adaptation index. Technical report, University of Notre Dame, Notre Dame, IN, 11, 2015.

[109] D. Meng-Chuen Chen, B. L. Bodirsky, T. Krueger, et al. The world's growing municipal solid waste: Trends and impacts. *Environmental Research Letters*, 15(074021), 2020.

[110] E. Chen, V. Narayanan, T. Pistochini, et al. Transient simultaneous heat and mass transfer model to estimate drying time in a wetted fur of a cow. *Biosystems Engineering*, 195:116–135, 2020.

[111] J. M. Chen and J. Liu. Evolution of evapotranspiration models using thermal and shortwave remote sensing data. *Remote Sensing of Environment*, 237(111594), 2020.

[112] L. Chen and P. A. Dirmeyer. Global observed and modelled impacts of irrigation on surface temperature. *International Journal of Climatology*, 39(5):2587–2600, 2019.

[113] X. M. Chen, A. R. Tait, and D. D. Kitts. Flavonoid composition of orange peel and its association with antioxidant and anti-inflammatory activities. *Food Chemistry*, 218:15–21, 2017.

[114] E. Choi, A. J. Rigden, N. Tangdamrongsub, et al. US crop yield losses from hydroclimatic hazards. *Environmental Research Letters*, 19(014005), 2024.

[115] S. Chripko, R. Msadek, E. Sanchez-Gomez, et al. Impact of reduced Arctic sea ice on northern hemisphere climate and weather in autumn and winter. *Journal of Climate*, 34:5847–5867, 7 2021.

[116] A. F. Cibils, R. J. Lira Fernandez, G. E. Oliva, et al. Is holistic management really saving Patagonian rangelands from degradation? A response to teague. *Rangelands*, 36:26–27, 2014.

[117] M. A. Clark, M. Springmann, J. Hill, et al. Multiple health and environmental impacts of foods. *Proceedings of the National Academy of Sciences*, 116:23357–23362, 2019.

[118] H. J. Coelho-Junior, R. Calvani, M. Tosato, et al. Protein intake and physical function in older adults: A systematic review and meta-analysis. *Ageing Research Reviews*, 81:101731, 2022.

[119] J. Cohen, J. A. Screen, J. C. Furtado, et al. Recent Arctic amplification and extreme mid-latitude weather. *Nature Geoscience*, 7:627–637, 2014.

[120] D. Coley, M. Howard, and M. Winter. Food miles: Time for a rethink? *British Food Journal*, 113:919–934, 2011.

[121] J. A. Conder, D. A. Wiens, and J. Morris. On the decompression melting structure at volcanic arcs and back-arc spreading centers. *Geophysical Research Letters*, 29(15):17–1–17–4, 2002.

[122] A. R. Contosta, K. A. Arndt, E. E. Campbell, et al. Management intensive grazing on New England dairy farms enhances soil nitrogen stocks and elevates soil nitrous oxide emissions without increasing soil carbon. *Agriculture, Ecosystems & Environment*, 317(107471), 2021.

[123] B. W. Conway-Jones and N. White. Paleogene buried landscapes and climatic aberrations triggered by mantle plume activity. *Earth and Planetary Science Letters*, 593(117644), 2022.

[124] A. A. Cook. Measuring methane emissions from American bison (bison bison l.) using eddy covariance. Master's thesis, Montana State University, Bozeman, College of Agriculture, 2019.

[125] B. I. Cook. *Drought: An Interdisciplinary Perspective*. Columbia University Press, 2019.

[126] B. I. Cook, K. J. Anchukaitis, R. Touchan, et al. Spatiotemporal drought variability in the mediterranean over the last 900 years. *Journal of Geophysical Research: Atmospheres*, 121(5):2060–2074, 2016.

[127] B. I. Cook, S. P. Shukla, M. J. Puma, et al. Irrigation as an historical climate forcing. *Climate Dynamics*, 44:1715–1730, 2015.

[128] A. Coquereau, F. Sévellec, T. Huck, et al. Anthropogenic changes of interannual-to-decadal climate variability in cmip6 multi-ensemble simulations. *Journal of Climate*, 2024.

[129] K. D. Corbin, E. A. Carnero, B. Dirks, et al. Host-diet-gut microbiome interactions influence human energy balance: A randomized clinical trial. *Nature Communications*, 14(3161), 2023.

[130] L. Cordain, J. B. Miller, S. Boyd Eaton, et al. Plant-animal subsistence ratios and macronutrient energy estimations in worldwide hunter-gatherer diets. *American Journal of Clinical Nutrition*, 71(3):682–692, 2000.

[131] E. Cottrell, S. K. Birner, M. Brounce, et al. Oxygen fugacity across tectonic settings. In Roberto Moretti and Daniel R. Neuville, editors, *Magma Redox Geochemistry*, Geophysical Monograph 266, First Edition. American Geophysical Union, John Wiley & Sons, Washington, DC, and Hoboken, NJ, 2022.

BIBLIOGRAPHY

[132] A. E. Cowan, R. L. Bailey, S. Jun, et al. The total nutrient index is a useful measure for assessing total micronutrient exposures among US adults. *Journal of Nutrition*, 152(3):863–871, 2022.

[133] M. Crippa, E. Solazzod, D. Guizzardi, et al. Food systems are responsible for a third of global anthropogenic GHG emissions. *Nature Food*, 2, 2021.

[134] A. Cristea, D. Hummels, L. Puzzello, et al. Trade and the greenhouse gas emissions from international freight transport. *Journal of Environmental Economics and Management*, 65(1):153–173, 2013.

[135] C. L. Curl, J. Porter, I. Penwell, et al. Effect of a 24-week randomized trial of an organic produce intervention on pyrethroid and organophosphate pesticide exposure among pregnant women. *Environment International*, 132:104957, 2019.

[136] D. F. Cusack, C. E. Kazanski, A. Hedgpeth, et al. Reducing climate impacts of beef production: A synthesis of life cycle assessments across management systems and global regions. *Global Change Biology*, 27:1721–1736, 2021.

[137] R. G. da Silva and A. S. C. Maia. Evaporative cooling and cutaneous surface temperature of Holstein cows in tropical conditions. *Revista Brasileira de Zootecnia*, 40:1143–1147, 2011.

[138] J. E. Dalen, J. S. Alpert, R. J. Goldberg, et al. The epidemic of the 20th century: Coronary heart disease. *American Journal of Medicine*, 127(9):807–812, 2014.

[139] C. C. de Melo Costa, A. S. C. Maia, et al. Thermal balance of nellore cattle. *International Journal of Biometeorology*, 62(5):723–731, 2018.

[140] X. D. de Otalora, L. Epelde, J. Arranz, et al. Regenerative rotational grazing management of dairy sheep increases springtime grass production and topsoil carbon storage. *Ecological Indicators*, 125(107484), 2021.

[141] T. de Ponti, B. Rijk, and M. K. van Ittersum. The crop yield gap between organic and conventional agriculture. *Agricultural Systems*, 108:1–9, 2012.

[142] R. J. de Souza, A. Mente, A. Maroleanu, et al. Intake of saturated and trans unsaturated fatty acids and risk of all cause mortality, cardiovascular disease, and type 2 diabetes: Systematic review and meta-analysis of observational studies. *British Medical Journal*, 351(h3978), 2015.

[143] A. DeAngelis, F. Dominguez, Y. Fan, et al. Evidence of enhanced precipitation due to irrigation over the great plains of the United States. *Journal of Geophysical Research Atmospheres*, 115(D15115), 2010.

[144] E. del Val, J. J. Armesto, O. Barbosa, et al. Rain forest islands in the Chilean semiarid region: Fog-dependency, ecosystem persistence and tree regeneration. *Ecosystems*, 9(598), 2006.

[145] F. Delaney. *Simple Courage: The True Story of Peril on the Sea*. Random House, New York, 2007.

[146] J. D. Derner, D. J. Augustine, D. D. Briske, et al. Can collaborative adaptive management improve cattle production in multipaddock grazing systems? *Rangeland Ecology and Management*, 75:1–8, 2021.

[147] S. J. Dettenmaier, T. A. Messmer, T. J. Hovick, et al. Effects of livestock grazing on rangeland biodiversity: A meta-analysis of grouse populations. *Ecology and Evolution*, 7:7620–7627, 2017.

[148] M. S. Di Bitetti, M. E. Iezzi, P. Cruz, et al. Effects of cattle on habitat use and diel activity of large native herbivores in a South American rangeland. *Journal for Nature Conservation*, 58:125900, 2020.

[149] F. G. Di Girolamo, R. Situlin, N. Fiotti, et al. Higher protein intake is associated with improved muscle strength in elite senior athletes. *Nutrition*, 42:82–86, 2017.

[150] J. Diamond. *Collapse: How Societies Choose to Fail or Succeed*. Penguin, London, and New York, 2004.

[151] S. Diaz, J. Settele, and E. S. Brondízio et al. Global assessment report on biodiversity and ecosystem services, et al. Online at https://www.ipbes.net/global-assessment; accessed 12 June 2021.

[152] S. Dikmen, F. A. Khan, H. J. Huson, et al. The slick hair locus derived from senepol cattle confers thermotolerance to intensively managed lactating Holstein cows. *Journal of Dairy Science*, 97(9):5508–5520, 2014.

[153] M. Dinu, R. Abbate, G. F. Gensini, et al. Vegetarian, vegan diets and multiple health outcomes: A systematic review with meta-analysis of observational studies. *Critical Reviews in Food Science and Nutrition*, 57(17):3640–3649, 2017.

[154] A. E. Disher, K. L. Stewart, A. J. E. Bach, et al. Contribution of dietary composition on water turnover rates in active and sedentary men. *Nutrients*, 13(2124):1–12, 2021.

[155] T. Dobzhansky. Nothing in biology makes sense except in the light of evolution. *American Biology Teacher*, 35(3):125–129, 1973.

[156] N. G. G. Domingo, S. Balasubramanian, S. K. Thakrar, et al. Air quality-related health damages of food. *Proceedings of the National Academy of Sciences*, 118(20), 2021.

[157] E. Donadio, S. Di Martino, and S. Heinonen. Rewilding Argentina: Lessons for the 2030 biodiversity targets. *Nature*, 603:225–227, 2022.

[158] M. Donovan. Modelling soil loss from surface erosion at high-resolution to better understand sources and drivers across land uses and catchments; a

national-scale assessment of Aotearoa, New Zealand. *Environmental Modelling & Software*, 147:105228, 2022.

[159] D. F. Dowling. Significance of sweating in heat tolerance of cattle. *Australian Journal of Agricultural Research*, 9:579–586, 1958.

[160] W. Duan, B. He, D. Nover, et al. Floods and associated socioeconomic damages in China over the last century. *Natural Hazards*, 82:401–413, 05 2016.

[161] N. Dudley and S. Alexander. Agriculture and biodiversity: A review. *Biodiversity*, 18(2-3):45–49, 2017.

[162] K. E. Dybala, V. Matzek, T. Gardali, et al. Carbon sequestration in riparian forests: A global synthesis and meta-analysis. *Global Change Biology*, 25:57–67, 2019.

[163] ECOINVENT. ecoinvent database.

[164] S. R. Edwards, J. D. Hobbs, and J. T. Mulliniks. High milk production decreases cow-calf productivity within a highly available feed resource environment. *Translational Animal Science*, 1:54–59, 2017.

[165] EESI. Fact sheet: Vehicle efficiency and emissions standards. 8 2015.

[166] T. Egan. *The Worst Hard Time: The Untold Story of Those Who Survived the Great American Dust Bowl*. Mariner Books, Boston and New York, reprint edition edition, 2006.

[167] J. Eilperin and B. Dennis. How James Inhofe is upending the nation's energy and environmental policies. *Washington Post*, 3, 2017.

[168] C. Einhorn and N. Popovich. This map shows where biodiversity is most at risk in America. *New York Times*, 3, 2022.

[169] L. T. Elkins-Tanton. Magma oceans in the inner solar system. *Annual Review of Earth and Planetary Science*, 40:113–39, 2012.

[170] R. W. Emerson. *Society and Solitude: Twelve Chapters; The Works of Ralph Waldo Emerson, vol. 7*. Classics Today, United States, 2022.

[171] P. Escobar-Bahamondes, M. Oba, and K. A. Beauchemin. Universally applicable methane prediction equations for beef cattle fed high- or low-forage diets. *Canadian Journal of Animal Science*, 97(1):83–94, 2017.

[172] G. Eshel. *Spatiotemporal Data Analysis*. Princeton University Press, Princeton NJ, Woodstock UK, Beijing, China, 2011.

[173] G. Eshel. How to prioritize voluntary dietary modification. *Advances in Environmental and Engineering Research*, 1:1–8, 2020.

[174] G. Eshel. Disproportionate contributions to air quality-related deaths: The latest case against red meat. *Proceedings of the National Academy of Sciences*, 118(22), 2021.

[175] G. Eshel. Small-scale integrated farming systems can abate continental-scale nutrient leakage. *PLOS Biology*, 19(e3001264), 2021.

[176] G. Eshel, A. Dayalu, S. C. Wofsy, et al. Listening to the forest: An artificial neural network-based model of carbon uptake at Harvard Forest. *Journal of Geophysical Research: Biogeosciences*, 124(3):461–478, 2019.

[177] G. Eshel, A. I. Flamholz, R. Milo, et al. US grass fed beef is not more carbon intensive than industrial beef, and is ≈10-fold more intensive than common protein equivalent alternatives. *Proceedings of the National Academy of Sciences*, submitted, 2024.

[178] G. Eshel and P. A. Martin. Diet, energy and global warming. *Earth Interactions*, 10(9):1–17, 2006.

[179] G. Eshel, A. Shepon, T. Makov, et al. Land, irrigation water, greenhouse gas, and reactive nitrogen burdens of meat, eggs, and dairy production in the United States. *Proceedings of the National Academy of Sciences*, 111(33):11996–12001, 2014.

[180] G. Eshel, A. Shepon, T. Makov, et al. Partitioning United States' feed consumption among livestock categories for improved environmental cost assessments. *Journal of Agricultural Science*, 153(3):432–445, 2014.

[181] G. Eshel, A. Shepon, E. Noor, et al. Environmentally optimal, nutritionally aware beef replacement plant-based diets. *Environmental Science & Technology*, 50:8164–8168, 2016.

[182] G. Eshel, A. Shepon, T. Shaket, et al. A model for "sustainable" US beef production. *Nature Ecology & Evolution*, 2:81–85, 2018.

[183] G. Eshel, P. Stainier, A. Shepon, et al. Environmentally optimal, nutritionally sound, protein and energy conserving plant based alternatives to US meat. *Scientific Reports*, 9:10345, 2019.

[184] C. Falcoa, M. Galeotti, and A. Olperac. Climate change and migration: Is agriculture the main channel? *Global Environmental Change*, 59(101995), 2019.

[185] M. Falkenmark, L. Wang-Erlandsson, and J. Rockström. Understanding of water resilience in the Anthropocene. *Journal of Hydrology X*, 2(100009), 2019.

[186] Y. Fang, S. Lee, H. Xu, et al. Organic controls over biomineral ca-mg carbonate compositions and morphologies. *Crystal Growth & Design*, 23:4872–4882, 2023.

[187] FAO. Fao statistical database, food balances 2010-, 2023. Accessed 28 February 2023.

[188] S. Farchi, M. De Sario, E. Lapucci, et al. Meat consumption reduction in italian regions: Health co-benefits and decreases in GHG emissions. *PLoS One*, 12(e0182960), 2017.

[189] B. F. Farrell. Equable climate dynamics. *Journal of the Atmospheric Sciences*, 47:2986–2995, 1990.

[190] M. S. Farvid, W. Y. Chen, B. A. Rosner, et al. Fruit and vegetable consumption and breast cancer incidence: Repeated measures over 30 years of follow-up. *International Journal of Cancer*, 144(7):1496–1510, 2019.

[191] M. R. Felipe-Lucia, S. Soliveres, C. Penone, et al. Land-use intensity alters networks between biodiversity, ecosystem functions, and services. *Proceedings of the National Academy of Sciences*, 117(45):28140–28149, 2020.

[192] T. R. Fenton and C. J. Fenton. Paleo diet still lacks evidence. *American Journal of Clinical Nutrition*, 104:844–844, 09, 2016.

[193] B. G. Ferguson, S. A. W. Diemzont, R. Alfaro-Arguello, et al. Sustainability of holistic and conventional cattle ranching in the seasonally dry tropics of Chiapas, Mexico. *Agricultural Systems*, 120:38–48, 2013.

[194] C. R. Ferguson. Changes in Great Plains low-level jet structure and associated precipitation over the 20th century. *Journal of Geophysical Research: Atmospheres*, 127(e2021JD035859), 2022.

[195] K. A. Fesenmyer and D. C. Dauwalter. Livestock management, beaver, and climate influences on riparian vegetation in a semi-arid landscape. *PLoS One*, 13(e0208928), 2018.

[196] R. E. Fewster, P. J. Morris, R. F. Ivanovic, et al. Imminent loss of climate space for permafrost peatlands in Europe and western Siberia. *Nature Climate Change*, 12:373–379, 2022.

[197] A. Filazzola, C. Brown, M. A. Dettlaff, et al. The effects of livestock grazing on biodiversity are multi-trophic: A meta-analysis. *Ecology Letters*, 23(8):1298–1309, 2020.

[198] C. G. Fischer and T. Garnett. Plates, pyramids, and planets. Technical report, Food Climate Research Network, 2016.

[199] R. A. Fischer, E. Cottrell, E. Hauri, et al. The carbon content of earth and its core. *Proceedings of the National Academy of Sciences*, 117(16):8743–8749, 2020.

[200] Food and Agriculture Organization. Land use. 2021.

[201] United Nations Food and Agricultural Organization, editors. *The State of Food Security and Nutrition in the World 2021*. Number 2021 in The State of Food Security and Nutrition in the World (SOFI). FAO, IFAD, UNICEF, WFP and WHO, 2021.

[202] National Centers for Environmental Information. Climate at a glance: Global time series, 2021. Online at https://www.ncdc.noaa.gov/cag/; accessed 31 May 2021.

[203] D. Foreman. *Rewilding North America: A Vision for Conservation in the 21st Century*. Island Press, Washington, DC, 2004.

[204] National Waterways Foundation.

[205] W. Fracz, G. Janowski, and G. Ryzinska. Selected aspects of manufacturing and strength evaluation of porous composites based on numerical simulations. *Mechanika*, 89:31–43, 2017.

[206] A. C. Franke and E. Kotze. High-density grazing in Southern Africa: Inspiration by nature leads to conservation? *Outlook on Agriculture*, 51:67–74, 2022.

[207] S. Friel, A. D. Dangour, T. Garnett, et al. Public health benefits of strategies to reduce greenhouse-gas emissions: Food and agriculture. *The Lancet*, 374(9706):2016–2025, 2009.

[208] S. Frith. *The Evidence for Holistic Planned Grazing*, chapter 5, pages 89–106. McGill-Queen's University Press, Montreal, Quebec, Canada, 2020.

[209] C. D. Fryar, M. D. Carroll, Q. Gu, et al. Anthropometric reference data for children and adults: United States, 2015–2018. Technical Report 3(46), National Center for Health Statistics, U.S. Department of Health and Human Services, Centers for Disease Control and Prevention, Hyattsville, MD, 1 2021. Vital Health Statistics, Series 3, No. 46, CS321244.

[210] S. Fujimori, W. Wu, J. Doelman, et al. Land-based climate change mitigation measures can affect agricultural markets and food security. *Nature Food*, 3:110–121, 2022.

[211] S. Gabici. Low-energy cosmic rays: Regulators of the dense interstellar medium. *Astronomy and Astrophysics Review*, 30(4), 2022.

[212] J. N. Galloway, J. D. Aber, J. Willem Erisman, et al. The nitrogen cascade. *BioScience*, 53:341–356, 2003.

[213] M. L. Galyean, K. A. Beauchemin, J. Caton, et al. *Nutrient Requirements of Beef Cattle: Eighth Revised Edition*. The National Academies Press, Washington, DC, 2016.

[214] A. C. Ganguli and M. E. O'Rourke. How vulnerable are rangelands to grazing? *Science*, 378(6622):834–834, 2022.

[215] C. D. Gardner, J. C. Hartle, R. D. Garrett, et al. Maximizing the intersection of human health and the health of the environment with regard to the amount and type of protein produced and consumed in the United States. *Nutrition Reviews*, 77(4):197–215, 02 2019.

[216] R. H. Gardner and K. A. M. Engelhardt. Spatial processes that maintain biodiversity in plant communities. *Perspectives in Plant Ecology, Evolution and Systematics*, 9:211–228, 2008.

BIBLIOGRAPHY   233

[217] G. Garland, S. Banerjee, A. Edlinger, et al. A closer look at the functions behind ecosystem multifunctionality: A review. *Journal of Ecology*, 109:600–613, 2021.

[218] G. F. Gause, O. K. Nastukova, and W. W. Alpatov. The influence of biologically conditioned media on the growth of a mixed population of paramecium caudatum and p. aureliax. *Journal of Animal Ecology*, 3:222–230, 1934.

[219] M. Gautam and M. Agrawal. *Greenhouse Gas Emissions from Municipal Solid Waste Management: A Review of Global Scenario*, chapter 8, pp. 112–134. Springer, Singapore, 2021.

[220] K. G. Gebremedhin, P. E. Hillman, C. N. Lee, et al. Sweating rates of dairy cows and beef heifers in hot conditions. *Transactions of the American Society of Agricultural and Biological Engineers*, 51(6):2167–2178, 2008.

[221] K. G. Gebremedhin and B. Wu. Simulation of sensible and latent heat losses from wet-skin surface and fur layer. *Journal of Thermal Biology*, 27:291–297, 2002.

[222] V. Geissen, V. Silva, E. H. Lwanga, et al. Cocktails of pesticide residues in conventional and organic farming systems in Europe – legacy of the past and turning point for the future. *Environmental Pollution*, 278:116827, 2021.

[223] R. Gelaro, W. McCarty, M. J. Suárez, et al. The modern-era retrospective analysis for research and applications, version 2 (merra-2). *Journal of Climate*, 30:5419–5454, 2017.

[224] K. S. Rawat, G. T. Patle, T. T. Sikar, et al. Estimation of infiltration rate from soil properties using regression model for cultivated land. *Geology, Ecology, and Landscapes*, 3(1):1–13, 2019.

[225] K. E. Giller, R. Hijbeek, J. A. Andersson, et al. Regenerative agriculture: An agronomic perspective. *Outlook on Agriculture*, 50:13–25, 2021.

[226] J. Gillespie. Milk cost of production estimates. Technical report, United States Department of Agriculture, Economic Research Service, Washington, DC, 5, 2023.

[227] D. S. Glazier. Beyond the '3/4-power law': Variation in the intra- and interspecific scaling of metabolic rate in animals. *Biological Reviews*, 80, 2005.

[228] C. M. Godde, I. J. M. de Boer, E. zu Ermgassen, et al. Soil carbon sequestration in grazing systems: Managing expectations. *Climatic Change*, 161:385–391, 2020.

[229] H. Charles, J. Godfray, P. Aveyard, T. Garnett, et al. Meat consumption, health, and the environment. *Science*, 361(eaam5324), 2018.

[230] E. Goldsmith and N. Hildyard. *The Social and Environmental Effects of Large Dams*. Wadebridge Ecological Centre, Wadebridge, Cornwall, 1984.

[231] D. Goodrich, I. Burns, C. Unkrich, et al. Kineros2/agwa: Model use, calibration, and validation. *Transactions of the ASABE*, 55:1561–1574, 7, 2012.

[232] A. Gorguç, C. Bircan, and F. M. Yılmaz. Sesame bran as an unexploited by-product: Effect of enzyme and ultrasound-assisted extraction on the recovery of protein and antioxidant compounds. *Food Chemistry*, 283:637–645, 2019.

[233] A. Gorguc, P. Ozer, and F. M. Yılmaz. Microwave-assisted enzymatic extraction of plant protein with antioxidant compounds from the food waste sesame bran: Comparative optimization study and identification of metabolomics using lc/q-tof/ms. *Journal of Food Processing and Preservation*, 44(1):e14304, 2020.

[234] J. R. Graff, N. B. Nelson, M. Roca-Martí, et al. Reconciliation of total particulate organic carbon and nitrogen measurements determined using contrasting methods in the North Pacific Ocean as part of the NASA EXPORTS field campaign. *Elementa: Science of the Anthropocene*, 11(00112), 12 2023.

[235] Y. T. Granja-Salcedo, R. M. Fernandes, R. C. de Araujo, et al. Long-term encapsulated nitrate supplementation modulates rumen microbial diversity and rumen fermentation to reduce methane emission in grazing steers. *Frontiers in Microbiology*, 10, 2019.

[236] T. R. Gregory. Evolution as fact, theory, and path. *Evolution: Education and Outreach*, 1:46–52, 2008.

[237] D. S. Grewal, R. Dasgupta, T. Hough, et al. Rates of protoplanetary accretion and differentiation set nitrogen budget of rocky planets. *Nature Geoscience*, 14:369–376, 2021.

[238] N. Grima, J. Brainard, and B. Fisher. Are wolves welcome? Hunters' attitudes towards wolves in Vermont, USA. *Oryx*, 55:262–267, 2021.

[239] M. Guasch-Ferre, A. Satija, S. A. Blondin, et al. Meta-analysis of randomized controlled trials of red meat consumption in comparison with various comparison diets on cardiovascular risk factors. *Circulation*, 139(15):1828–1845, 2019.

[240] K. E. Hales, T. M. Brown-Brandl, and H. C. Freetly. Effects of decreased dietary roughage concentration on energy metabolism and nutrient balance in finishing beef cattle. *Journal of Animal Science*, 92:264–271, 2014.

[241] A. Hallam. *A Revolution in the Earth Sciences: From Continental Drift to Plate Tectonics*. Clarendon Press, Oxford, UK, 1973.

[242] A. N. Halliday and R. M. Canup. The accretion of planet Earth. *Nature Reviews Earth & Environment*, 4:19–35, 2023.

[243] C. Hankel and E. Tziperman. The role of atmospheric feedbacks in abrupt winter Arctic sea ice loss in future warming scenarios. *Journal of Climate*, 34:4435–4447, 2021.

BIBLIOGRAPHY 235

[244] X. Hao, J. Yang, S. Dong, et al. The influence of grazing intensity on soil organic carbon storage in grassland of China: A meta-analysis. *Science of the Total Environment*, 924(171439), 2024.

[245] G. Hardin. The competitive exclusion principle. *Science*, 131:1292–1297, 1960.

[246] K. Hardy, J. Brand-Miller, K. D. Brown, et al. The importance of dietary carbohydrate in human evolution. *Quarterly Review of Biology*, 90(3):251–268, 2015.

[247] W. D. Harkins. The evolution of the elements and the stability of complex atoms. i. a new periodic system which shows a relation between the abundance of the elements and the structure of the nuclei of atoms. *Journal of the American Chemial Society*, 39:856–879, 1917.

[248] K. R. Harmoney and J. R. Jaeger. Using modified intensive early stocking for grazing replacement heifers. *Kansas Agricultural Experiment Station Research Reports*, 4, 2018.

[249] G. Hasha. Livestock feeding and feed imports in the European Union—A decade of change. Technical Report FDS-0602-01, United States Department of Agriculture, Electronic Outlook Report from the Economic Research Service, 7, 2002.

[250] M. Hashimoto, K. Matsuzaki, K. Maruyama, et al. Perilla seed oil in combination with nobiletin-rich ponkan powder enhances cognitive function in healthy elderly Japanese individuals: A possible supplement for brain health in the elderly. *Food Function*, 13:2768–2781, 2022.

[251] H.-J. Hawkins, Z.-S. Venter, and M. D. Cramer. A holistic view of holistic management: What do farm-scale, carbon, and social studies tell us? *Agriculture, Ecosystems & Environment*, 323(107702), 2022.

[252] M. N. Hayek and R. D. Garrett. Nationwide shift to grass-fed beef requires larger cattle population. *Environmental Research Letters*, 13(084005), 2018.

[253] R. M. Hazen. *The Story of Earth: The First 4.5 Billion Years, from Stardust to Living Planet*. Penguin Books, New York, 2013.

[254] T. Van Hecke, L. M. A. Jakobsen, E. Vossen, et al. Short-term beef consumption promotes systemic oxidative stress, TMAO formation and inflammation in rats, and dietary fat content modulates these effects. *Food & Function*, 7:3760–3771, 2016.

[255] M. Heid. Experts say lobbying skewed the US dietary guidelines. 1, 2016.

[256] C. Heinze, T. Blenckner, H. Martins, et al. The quiet crossing of ocean tipping points. *Proceedings of the National Academy of Sciences*, 118(e2008478118), 2021.

[257] E. Hemingway. *Islands in the Stream*. Scribner's, New York, 1970.

[258] B. B. Henderson, P. J. Gerber, T. E. Hilinski, et al. Greenhouse gas mitigation potential of the world's grazing lands: Modeling soil carbon and nitrogen fluxes of mitigation practices. *Agriculture, Ecosystems & Environment*, 207:91–100, 2015.

[259] B. Henry. Potential for soil carbon sequestration in northern Australian grazing lands: A review of the evidence. Technical report, Department of Agriculture and Fisheries, Queensland, 2023 Project AS10309_6452.

[260] A. J. Hill, T. E. Dawson, O. Shelef, et al. The role of dew in Negev Desert plants. *Oecologia*, 178:317–327, 2015.

[261] N. T. Hoang, O. Taherzadeh, H. Ohashi, et al. Mapping potential conflicts between global agriculture and terrestrial conservation. *Proceedings of the National Academy of Sciences*, 120(e2208376120), 2023.

[262] J. M. Hodgson, V. Burke, L. J. Beilin, et al. Partial substitution of carbohydrate intake with protein intake from lean red meat lowers blood pressure in hypertensive persons. *American Journal of Clinical Nutrition*, 83(4):780–787, 06 2006.

[263] R. M. Holdo, R. D. Holt, M. B. Coughenour, et al. Plant productivity and soil nitrogen as a function of grazing, migration and fire in an African savanna. *Journal of Ecology*, 95:115–128, 2007.

[264] A. F. Holt and C. B. Condit. Slab temperature evolution over the lifetime of a subduction zone. *Geochemistry, Geophysics, Geosystems*, 22(e2020GC009476), 2021.

[265] D. L. Hoover, O. L. Hajek, M. D. Smith, et al. Compound hydroclimatic extremes in a semi-arid grassland: Drought, deluge, and the carbon cycle. *Global Change Biology*, 28(8):2611–2621, 2022.

[266] P. Hopcroft. Ancient ice and the global methane cycle. *Nature*, 548:403–404, 2017.

[267] H. Hoppeler and E. R. Weibel. Scaling functions to body size: Theories and facts. *Journal of Experimental Biology*, 208:1573–1574, 2005.

[268] S. R. Hoy, J. A. Vucetich, and R. O. Peterson. The role of wolves in regulating a chronic non-communicable disease, osteoarthritis, in prey populations. *Frontiers in Ecology and Evolution*, 10, 2022.

[269] R. M. Hughes and R. L. Vadas. Agricultural effects on streams and rivers: A western USA focus. *Water*, 13(1901), 2021.

[270] M. P. Huijser, P. McGowen, J. Fuller, et al. Wildlife-vehicle collision reduction study: Report to congress. Report FHWA-HRT-08-034, Federal Highway Administration, Office of Safety Research and Development, McLean, VA, 8 2008.

[271] M. Hulme. Is it too late (to stop dangerous climate change)? An editorial. *WIREs Climate Change*, 11(e619), 2020.

[272] D. Hunter, M. Foster, J. O. McArthur, et al. Evaluation of the micronutrient composition of plant foods produced by organic and conventional agricultural methods. *Critical Reviews in Food Science and Nutrition*, 51:571–582, 2011.

[273] D. A. Hutchins and D. G. Capone. The marine nitrogen cycle: New developments and global change. *Nature Reviews Microbiology*, 20:401–414, 2022.

[274] I. Iguacel, M. L. Miguel-Berges, A. Gomez-Bruton, et al. Veganism, vegetarianism, bone mineral density, and fracture risk: A systematic review and meta-analysis. *Nutrition Reviews*, 77:1–18, 2019.

[275] O. C. Doering III, J. N. Galloway, T. L. Theis, et al. Reactive nitrogen in the United States: An analysis of inputs, flows, consequences, and management options: A report of the EPA science advisory board. Report EPA-SAB-11-013, U.S. Environmental Protection Agency, Science Advisory Board, Washington, DC, 8, 2011.

[276] Synthesis report of the IPCC sixth assessment report (ar6), 2023.

[277] H. R. Isaacson, F. C. Hinds, M. P. Bryant, et al. Efficiency of energy utilization by mixed rumen bacteria in continuous culture. *Journal of Dairy Science*, 58:1645–1659, 1975.

[278] M. Ismail and T. Al-Ansari. Enhancing sustainability through resource efficiency in beef production systems using a sliding time window-based approach and frame scores. *Heliyon*, 9(e17773), 2023.

[279] S. Jafari, E. Hezaveh, Y. Jalilpiran, et al. Plant-based diets and risk of disease mortality: A systematic review and meta-analysis of cohort studies. *Critical Reviews in Food Science and Nutrition*, 0(0):1–13, 2021.

[280] M. Jain, C. B. Barrett, D. Solomon, et al. Surveying the evidence on sustainable intensification strategies for smallholder agricultural systems. *Annual Review of Environment and Resources*, 48(1):347–369, 2023.

[281] M. Janowiak, W. J. Connelly, K. Dante-Wood, et al. Considering forest and grassland carbon in land management. Technical Report General Technical Report WO-95, United States Department of Agriculture, United States Forest Service, Madison, WI, 6, 2017.

[282] E. A. Jara and C. K. Winter. Safety levels for organophosphate pesticide residues on fruits, vegetables, and nuts. *International Journal of Food Contamination*, 6(6), 2019.

[283] L. Jarosz and J. Qazi. The geography of Washington's world apple: Global expressions in a local landscape. *Journal of Rural Studies*, 16(1):1–11, 2000.

[284] K. Jaster. Ditching meat isn't the answer for climate change. Better farming is, 5, 2021. Online at https://www.washingtonpost.com/outlook/ditching-meat-isnt-the-answer-for-climate-change-better-farming-is/2021/05/14/86001c36-b426-11eb-ab43-bebddc5a0f65_story.html; accessed 23 June 2021.

[285] C. N. Jenkins, K. S. Van Houtan, S. L. Pimm, et al. US protected lands mismatch biodiversity priorities. *Proceedings of the National Academy of Sciences*, 112:5081–5086, 2015.

[286] J.-P. Jenny, S. Koirala, I. Gregory-Eaves, et al. Human and climate global-scale imprint on sediment transfer during the holocene. *Proceedings of the National Academy of Sciences*, 116(46):22972–22976, 2019.

[287] P. Jepson and C. Blythe. *Rewilding: The Radical New Science of Ecological Recovery*. Icon Books, London, England, 2020.

[288] A. R. Johnson. The paleo diet and the American weight loss utopia, 1975–2014. *Utopian Studies*, 26:101–124, 2015.

[289] B. Johnson and C. Goldblatt. The nitrogen budget of Earth. *Earth-Science Reviews*, 148:150–173, 2015.

[290] B. C. Johnston, D. Z. abd Mi Ah Han, R. W. M. Vernooij, et al. Unprocessed red meat and processed meat consumption: Dietary guideline recommendations from the nutritional recommendations (nutrirecs) consortium. *Annals of Internal Medicine*, 171(10):756–764, 2019.

[291] J. Jordahl, M. McDaniel, B. A. Miller, et al. Carbon storage in cropland soils: Insights from iowa, United States. *Land*, 12(1630), 2023.

[292] T. R. Jorns, J. D. Scasta, J. D. Derner, et al. Adaptive multi-paddock grazing management reduces diet quality of yearling cattle in shortgrass steppe. *Rangeland Journal*, 45:160–172, 2023.

[293] S. Jurek. *Eat & Run: My Unlikely Journey to Ultramarathon Greatness*. Houghton Mifflin Harcourt, Boston and New York, 2012.

[294] S. Kadoya, D. C. Catling, R. W. Nicklas, et al. Mantle data imply a decline of oxidizable volcanic gases could have triggered the great oxidation. *Nature Communications*, 11(2774), 2020.

[295] J. Kaiser. Wounding Earth's fragile skin. *Science*, 304:1616–1618, 2004.

[296] C. Kaleta, S. Schauble, U. Rinas, et al. Metabolic costs of amino acid and protein production in Escherichia coli. *Biotechnology Journal*, 8:1105–1114, 2013.

[297] S. Kamerlin, P. Sharma, R. Prasad, et al. Why nature really chose phosphate. *Quarterly Reviews of Biophysics*, 46:1–132, 2013.

[298] M. Kanamitsu, W. Ebisuzaki, J. Woollen, et al. NCEO-DOE AMIP-II reanalysis (r-2). *Bulletin of the American Meteorological Society*, 83:1631–1643, 2002.

BIBLIOGRAPHY

[299] L. Kantor and A. Blazejczyk. Food availability (per capita) data system, 2021. Online at https://www.ers.usda.gov/data-products/food-availability-per-capita-data-system/food-availability-per-capita-data-system/; accessed 23 June 2021.

[300] J. O. Kaplan, K. M. Krumhardt, M.-J. Gaillard, et al. Constraining the deforestation history of Europe: Evaluation of historical land use scenarios with pollen-based land cover reconstructions. *Land*, 6(4):91, 2017.

[301] A. B. Kara, P. A. Rochford, and H. E. Hurlburt. Mixed layer depth variability over the global ocean. *Journal of Geophysical Research: Oceans*, 108(C3):1–15, 2003.

[302] D. L. Katz. *The Truth About Food: Why Pandas Eat Bamboo and People Get Bamboozled.* Independently published, New Haven, CT, 2018.

[303] D. L. Katz, K. N. Doughty, K. Geagan, et al. Perspective: The public health case for modernizing the definition of protein quality. *Advances in Nutrition*, 10(5):755–764, 2019.

[304] D. L. Katz, E. P. Frates, J. P. Bonnet, et al. Lifestyle as medicine: The case for a true health initiative. *American Journal of Health Promotion*, 32:1452–1458, 2018.

[305] R. Kazimierczak, D. Srednicka-Tober, J. Golba, et al. Evaluation of pesticide residues occurrence in random samples of organic fruits and vegetables marketed in Poland. *Foods*, 11(13), 2022.

[306] C. P. Kelley, S. Mohtadi, M. A. Cane, et al. Climate change in the fertile crescent and implications of the recent syrian drought. *Proceedings of the National Academy of Sciences*, 112:3241–3246, 2015.

[307] F. M. Kelliher and H. Clark. Methane emissions from bison—an historic herd estimate for the North American Great Plains. *Agricultural and Forest Meteorology*, 150:473–477, 2010.

[308] G. Keppel-Aleks and R. A. Washenfelder. The effect of atmospheric sulfate reductions on diffuse radiation and photosynthesis in the United States during 1995–2013. *Geophysical Research Letters*, 43:9984–9993, 2016.

[309] C. Ketcham. *This Land: How Cowboys, Capitalism, and Corruption are Ruining the American West.* Viking, New York, 2019.

[310] L. Khalil and J. Roose. Anti-government extremism in Australia: Understanding the Australian anti-lockdown freedom movement as a complex anti-government social movement. *Perspectives on Terrorism*, 17:144–169, 2023.

[311] U. Khatri-Chhetri, S. Banerjee, K. A. Thompson, et al. Cattle grazing management affects soil microbial diversity and community network complexity in the Northern Great Plains. *Science of the Total Environment*, 912(169353), 2024.

[312] U. Khatri-Chhetri, K. A. Thompson, S. A. Quideau, et al. Adaptive multi-paddock grazing increases mineral associated soil carbon in Northern grasslands. *Agriculture, Ecosystems & Environment*, 369(109000), 2024.

[313] B. H. Kim and G. M. Gadd. *Bacterial Physiology and Metabolism*. Cambridge University Press, Cambridge, and New York, 1st edition, 2008.

[314] J. J. Kim, S. Ale, U. P. Kreuter, et al. Evaluating the impacts of alternative grazing management practices on soil carbon sequestration and soil health indicators. *Agriculture, Ecosystems & Environment*, 342:108234, 2023.

[315] S.-Ki Kim, H.-J. Kim, H. A. Dijkstra, et al. Slow and soft passage through tipping point of the atlantic meridional overturning circulation in a changing climate. *NPJ Climate and Atmospheric Science*, 5(13), 2022.

[316] S.-Ki Kim, J. Shin, S.-Il An, et al. Widespread irreversible changes in surface temperature and precipitation in response to CO_2 forcing. *Nature Climate Change*, 12:834–840, 2022.

[317] D. E. King, A. G. Mainous III, and C. A. Lambourne. Trends in dietary fiber intake in the United States, 1999–2008. *Journal of the Academy of Nutertion and Dietetics*, 112:642–648, 2012.

[318] H. Kirchmann. Why organic farming is not the way forward. *Outlook on Agriculture*, 48:22–27, 2019.

[319] K. K. Goldewijk, A. Beusen, G. van Drecht, et al. The hyde 3.1 spatially explicit database of human-induced global land-use change over the past 12,000 years. *Global Ecology and Biogeography*, 20:73–86, 2011.

[320] E. Klinenberg. *Heat Wave: A Social Autopsy of Disaster in Chicago*. University Of Chicago Press, Chicago, 1st edition, 2002.

[321] S. T. Klopfenstein, D. R. Hirmas, and W. C. Johnson. Relationships between soil organic carbon and precipitation along a climosequence in loess-derived soils of the Central Great Plains, USA. *Catena*, 133:25–34, 2015.

[322] A. A. Koronowicz and P. Banks. Antitumor properties of CLA-enriched food products. *Nutrition and Cancer*, 70:529–545, 2018.

[323] A. N. Kravchenko, S. S. Snapp, and G. P. Robertson. Field-scale experiments reveal persistent yield gaps in low-input and organic cropping systems. *Proceedings of the National Academy of Sciences*, 114(5):926–931, 2017.

[324] C. J. Kruse, J. E. Warner, and L. E. Olson. A modal comparison of domestic freight transportation effects on the general public: 2001–2014. Technical report, Texas A&M Transportation Institute, The Texas A&M University System, College Station, TX, 2017.

[325] T. S. Kuhn. *The Structure of Scientific Revolutions*. University of Chicago Press, Chicago, 1962.

[326] L. Kunyeit, R. P. Rao, and K. A. Anu-Appaiah. Yeasts originating from fermented foods, their potential as probiotics and therapeutic implication for human health and disease. *Critical Reviews in Food Science and Nutrition*, 63:1–12, 2023.

[327] R. Lal. Digging deeper: A holistic perspective of factors affecting soil organic carbon sequestration in agroecosystems. *Global Change Biology*, 24:3285–3301, 2018.

[328] R. Lal. Regenerative agriculture for food and climate. *Journal of Soil and Water Conservation*, 75:123A–124A, 2020.

[329] R. Lal. Negative emission farming. *Journal of Soil and Water Conservation*, 76:61A–64A, 2021.

[330] K. Laland, B. Matthews, and M. W. Feldman. An introduction to niche construction theory. *Evolutionary Ecology*, 30:191–202, 2016.

[331] N. Lane. *Vital Question: Energy, Evolution, and the Origins of Complex Life*. W. W. Norton & Company, New York, and London, 2016.

[332] C. H. Langmuir and W. Broecker. *How to Build a Habitable Planet: The Story of Earth from the Big Bang to Humankind*. Princeton University Press, Princeton, NJ, Woodstock, UK, Beijing, China, 2012.

[333] F. M. Lappé. *Diet for a Small Planet*. Ballantine Books, New York, 1971.

[334] I. J. Larsen, P. C. Almond, A. Eger, et al. Rapid soil production and weathering in the Southern Alps, New Zealand. *Science*, 343(6171):637–640, 2014.

[335] Y. L. Cozler, C. Allain, C. Xavier, et al. Volume and surface area of Holstein dairy cows calculated from complete 3D shapes acquired using a high-precision scanning system: Interest for body weight estimation. *Computers and Electronics in Agriculture*, 165(104977), 2019.

[336] H. Lee, D. P. Schrag, M. Bunn, et al. Cambridge University Press, Cambridge, UK, and New York, 2021.

[337] N. K. Leeuwendaal, C. Stanton, P. W. O'Toole, et al. Fermented foods, health and the gut microbiome. *Nutrients*, 14(1527):1–26, 2022.

[338] C. Lefevre, F. Rekik, V. Alcantara, et al. Soil organic carbon: The hidden potential. Technical report, Food and Agriculture Organization of the United Nations, Rome, Italy, 2017.

[339] T. M. Lenton, T. W. Dahl, S. J. Daines, et al. Earliest land plants created modern levels of atmospheric oxygen. *Proceedings of the National Academy of Sciences*, 113(35):9704–9709, 2016.

[340] J. Leshy. America's public lands: Sketch of their political history and future challenges. *Natural Resources Journal*, 341–360, 2022.

[341] S. Lewandowsky, U. K. H. Ecker, and J. Cook. Beyond misinformation: Understanding and coping with the "post-truth" era. *Journal of Applied Research in Memory and Cognition*, 6:353–369, 2017.

[342] A.-M. Lezine, C. Robert, S. Cleuziou, et al. Climate change and human occupation in the Southern Arabian lowlands during the last deglaciation and the Holocene. *Global and Planetary Change*, 72(4):412–428, 2010.

[343] C. Li, L. M. Fultz, J. Moore-Kucera, et al. Soil carbon sequestration potential in semi-arid grasslands in the conservation reserve program. *Geoderma*, 294:80–90, 2017.

[344] J. Li, V. Narayanan, E. Kebreab, et al. A mechanistic thermal balance model of dairy cattle. *Biosystems Engineering*, 209:256–270, 2021.

[345] S. Li, T. Xing, R. Sa, et al. Effects of grazing on soil respiration in global grassland ecosystems. *Soil and Tillage Research*, 238(106033), 2024.

[346] J. S. Lin, M. V. M. Sarto, T. L. Carter, et al. Soil organic carbon, aggregation and fungi community after 44 years of no-till and cropping systems in the Central Great Plains, USA. *Archives of Microbiology*, 205(84), 2023.

[347] A. G. Liu, N. A. Ford, F. B. Hu, et al. A healthy approach to dietary fats: Understanding the science and taking action to reduce consumer confusion. *Nutrition Journal*, 16, 2017.

[348] N. Liu, X. Li, P. Zhao, et al. A review of chemical constituents and health-promoting effects of citrus peels. *Food Chemistry*, 365(130585), 2021.

[349] S. Liu, J. Liu, C. J. Wu, et al. Baseline and projected future carbon storage, carbon sequestration, and greenhouse-gas fluxes in terrestrial ecosystems of the eastern United States. Technical Report Professional Paper 1804, U.S. Department of the Interior, U.S. Geological Survey, Reston, VA, 2014. In: Zhu, Z. and Reed, B. C., eds, Baseline and projected future carbon storage and greenhouse-gas fluxes in ecosystems of the eastern United States.

[350] S. Liu, Y. Wu, C. J. Young, et al. Projected future carbon storage and greenhouse-gas fluxes of terrestrial ecosystems in the western United States. Technical Report Professional Paper 1797, U.S. Department of the Interior, U.S. Geological Survey, Reston, VA, 2012. In: Zhu, Z. and Reed, B. C., eds, Baseline and projected future carbon storage and greenhouse-gas fluxes in ecosystems of the western United States.

[351] R. F. Logan. *Causes, Climates and Distribution of Deserts*, volume 1, pp. 21–50. Academic Press, New York, and London, 1968.

[352] C. Lu, J. Zhang, B. Yi, et al. Riverine nitrogen footprint of agriculture in the Mississippi–Atchafalaya river basin: Do we trade water quality for crop production? *Environmental Research Letters*, 18(114043), 2023.

BIBLIOGRAPHY

[353] J. Lu, G. J. Carbone, X. Huang, et al. Mapping the sensitivity of agriculture to drought and estimating the effect of irrigation in the United States, 1950–2016. *Agricultural and Forest Meteorology*, 292-293(108124), 2020.

[354] J. Lynch and R. Pierrehumbert. Climate impacts of cultured meat and beef cattle. *Frontiers in Sustainable Food Systems*, 3, 2019.

[355] M. P. Lynch. Arrogance, truth and public discourse. *Episteme*, 15(3):283–296, 2018.

[356] T. Ma, B. V. Rolett, Z. Zheng, et al. Holocene coastal evolution preceded the expansion of paddy field rice farming. *Proceedings of the National Academy of Sciences*, 117(39):24138–24143, 2020.

[357] B. Machovina, K. J. Feeley, and W. J. Ripple. Biodiversity conservation: The key is reducing meat consumption. *Science of The Total Environment*, 536:419–431, 2015.

[358] E. Macia-Barber. *The Chemical Evolution of Phosphorus*. Apple Academic Press, Palm Bay, FL, 12, 2019.

[359] F. T. Maestre, Y. Le Bagousse-Pinguet, M. Delgado-Baquerizo, et al. Grazing and ecosystem service delivery in global drylands. *Science*, 378(6622):915–920, 2022.

[360] K. C. Makris, C. Konstantinou, X. D. Andrianou, et al. A cluster-randomized crossover trial of organic diet impact on biomarkers of exposure to pesticides and biomarkers of oxidative stress/inflammation in primary school children. *PLOS ONE*, 14(9):1–15, 09 2019.

[361] D. Malerba. The effects of social protection and social cohesion on the acceptability of climate change mitigation policies: What do we (not) know in the context of low- and middle-income countries? *European Journal of Development Research volume*, 34:1358–1382, 2022.

[362] R. Manavalan. A review about computational methods dedicated to apple fruit diseases detection. *Asian Research Journal of Current Science*, 3(ARJOCS.531):119–130, 2021.

[363] S. M. Manemann, Y. Gerber, S. J. Bielinski, et al. Recent trends in cardiovascular disease deaths: A state specific perspective. *BMC Public Health*, 21(1031), 2021.

[364] E. W. Manheimer. Reply to TR Fenton and CJ Fenton. *American Journal of Clinical Nutrition*, 104(3):845–845, 2016.

[365] J. S. Mankin, R. Seager, J. E. Smerdon, et al. Mid-latitude freshwater availability reduced by projected vegetation responses to climate change. *Nature Geosciences*, 12:983–988, 2019.

[366] F. Mariotti and C. D. Gardner. Dietary protein and amino acids in vegetarian diets—a review. *Nutrients*, 11(2661), 2019.

[367] F. W. Marlowe, J. C. Berbesque, B. Wood, et al. Honey, hadza, hunter-gatherers, and human evolution. *Journal of Human Evolution*, 71:119–128, 2014.

[368] K. A. Marsh, E. A. Munn, and S. K. Baines. Protein and vegetarian diets. *Medical Journal of Australia*, 199:S7–S10, 2013.

[369] S. J. Marshall. Regime shifts in glacier and ice sheet response to climate change: Examples from the northern hemisphere. In C. Adler, C. Derksen, M. Collins, and Z. Sebesvari, eds, *Knowledge Gaps From the IPCC Special Report on the Ocean and Cryosphere in a Changing Climate and Recent Advances*, Frontiers in Climate. Frontiers Media SA, 2022.

[370] M. Martinez, R. Yanez, J. Luis Alonso, et al. Chemical production of pectic oligosaccharides from orange peel wastes. *Industrial & Engineering Chemistry Research*, 49:8470–8476, 2010.

[371] B. Marty and N. Dauphas. The nitrogen record of crust-mantle interaction and mantle convection from archean to present. *Earth and Planetary Science Letters*, 206:397–410, 2003.

[372] B. Marty. Nitrogen content of the mantle inferred from N2-Ar correlation in oceanic basalts. *Nature*, 377:326–329, 1995.

[373] B. Marty. The origins and concentrations of water, carbon, nitrogen and noble gases on Earth. *Earth and Planetary Science Letters*, 313–314:56–66, 2012.

[374] P. Mason and T. Lang. *Sustainable Diets: How Ecological Nutrition Can Transform Consumption and the Food System*. Routledge, Taylor & Francis, London and New York, 2017.

[375] A. Matsakas and K. Patel. *Aerobic Metabolism*, pp. 29–32. Springer Berlin Heidelberg, Berlin, Heidelberg, 2012.

[376] K. Matsuzaki, A. Nakajima, Y. Guo, et al. A narrative review of the effects of citrus peels and extracts on human brain health and metabolism. *Nutrients*, 14, 2022.

[377] D. Mayerfeld. *The Limits of Grass*, pp. 157–175. Springer Food and Health (FH) Series, Cham, 2023.

[378] B. McCarthy, R. Anex, Y. Wang, et al. Trends in water use, energy consumption, and carbon emissions from irrigation: Role of shifting technologies and energy sources. *Environmental Science & Technology*, 54(23):15329–15337, 2020.

[379] K. McCoy, S. Peirce, G. Baker, et al. The importance of highways to US agriculture. Technical Report USDA AMS TM TSD, TS295.12-20, U.S. Department of Agriculture, Agricultural Marketing Service, Washington, DC, December 2020.

[380] C. McEvedy and R. Jones. *Atlas of World Population History*. Penguin, New York, 1978.

[381] M. McGee, C. Lenehan, P. Crosson, et al. Performance, meat quality, profitability, and greenhouse gas emissions of suckler bulls from pasture-based compared to an indoor high-concentrate weanling-to-beef finishing system. *Agricultural Systems*, 198(103379), 2022.

[382] M. D. McKenna, S. E. Grams, M. Barasha, et al. Organic and inorganic soil carbon in a semi-arid rangeland is primarily related to abiotic factors and not livestock grazing. *Geoderma*, 419(115844), 2022.

[383] R. G. McMurray, J. Soares, C. J. Caspersen, et al. Examining variations of resting metabolic rate of adults: A public health perspective. *Medicine & Science in Sports & Exercise*, 46:1352–1358, 2014.

[384] B. K. McNab. Standard energetics of phyllostomid bats: The inadequacies of phylogenetic-contrast analyses. *Comparative Biochemistry and Physiology, Part A: Molecular & Integrative Physiology*, 135:357–368, 2003.

[385] M. E. McSherry and M. E. Ritchie. Effects of grazing on grassland soil carbon: A global review. *Global Change Biology*, 19:1347–1357, 2013.

[386] E. Medawar, S. Huhn, A. Villringer, et al. The effects of plant-based diets on the body and the brain: A systematic review. *Translational Psychiatry*, 9(226), 2019.

[387] R. S. Meena, S. Kumar, R. Datta, et al. Impact of agrochemicals on soil microbiota and management: A review. *Land*, 9(2):34, 2020.

[388] V. Melina, W. Craig, and S. Levin. Position of the academy of nutrition and dietetics: Vegetarian diets. *Journal of the Academy of Nutrition and Dietetics*, 116:1970–1980, 2016.

[389] R. P. Mensink. Metabolic and health effects of isomeric fatty acids. *Current Opinion in Lipidology*, 16:27–30, 2005.

[390] R. Mesnage, I. N. Tsakiris, M. N. Antoniou, et al. Limitations in the evidential basis supporting health benefits from a decreased exposure to pesticides through organic food consumption. *Current Opinion in Toxicology*, 19:50–55, 2020. Mechanistic Toxicology.

[391] F. Michel, C. Hartmann, and M. Siegrist. Consumers' associations, perceptions and acceptance of meat and plant-based meat alternatives. *Food Quality and Preference*, 87:104063, 2021.

[392] A. J. Michels, J. A. Butler, S. L. Uesugi, et al. Multivitamin/multimineral supplementation prevents or reverses decline in vitamin biomarkers and cellular energy metabolism in healthy older men: A randomized, double-blind, placebo-controlled study. *Nutrients*, 15(12), 2023.

[393] C. Middleton. Lead-208 nuclei have thick skins. *Physics Today*, 74:12–14, 2021.

[394] M. M. Mihaylova, A. Chaix, M. Delibegovic, et al. When a calorie is not just a calorie: Diet quality and timing as mediators of metabolism and healthy aging. *Cell Metabolism*, 35:1114–1131, 2023.

[395] S. Mikhail and D. Sverjensky. Nitrogen speciation in upper mantle fluids and the origin of Earth's nitrogen-rich atmosphere. *Nature Geoscience*, 7:816–819, 2014.

[396] R. Milo, P. Jorgensen, U. Moran, et al. Bionumbers–the database of key numbers in molecular and cell biology. *Nucleic Acids Research*, 38(D750-3), 2010.

[397] S. M. Mingyang, T. T. Fung, F. B. Hu, et al. Association of animal and plant protein intake with all-cause and cause-specific mortality. *JAMA Internal Medicine*, 176:1453–1463, 2016.

[398] E. C. Mitchell, T. P. Fischer, D. R. Hilton, et al. Nitrogen sources and recycling at subduction zones: Insights from the Izu-Bonin-Mariana arc. *Geochemistry, Geophysics, Geosystems*, 11(2), 2010.

[399] K. Mitura, G. Cacak-Pietrzak, B. Feledyn-Szewczyk, et al. Yield and grain quality of common wheat (triticum aestivum l.) depending on the different farming systems (organic vs. integrated vs. conventional). *Plants*, 12(5), 2023.

[400] P. Mohammadpour and C. Grady. Regional analysis of nitrogen flow within the Chesapeake Bay watershed food production chain inclusive of trade. *Environmental Science & Technology*, 57(11):4619–4631, 2023.

[401] E. A. Mohareb, M. C. Heller, and P. M. Guthrie. Cities' role in mitigating United States food system greenhouse gas emissions. *Environmental Science and Technology*, 52(10):5545–5554, 2018.

[402] K. Mokany, S. Ferrier, T. D. Harwood, et al. Reconciling global priorities for conserving biodiversity habitat. *Proceedings of the National Academy of Sciences*, 117(18):9906–9911, 2020.

[403] A. Molotoks, P. Smith, and T. P. Dawson. Impacts of land use, population, and climate change on global food security. *Food and Energy Security*, 10(e261), 2021.

[404] G. Monbiot. *Regenesis: Feeding the World Without Devouring the Planet*. Penguin, London and New York, 2022.

[405] N. Monjotin, M. Josèphe Amiot, J. Fleurentin, et al. Clinical evidence of the benefits of phytonutrients in human healthcare. *Nutrients*, 14(1712), 2022.

[406] D. R. Montgomery. Soil erosion and agricultural sustainability. *Proceedings of the National Academy of Sciences*, 104(33):13268–13272, 2007.

[407] A. Moodie. Before you read another health study, check who's funding the research, 12, 2016.

BIBLIOGRAPHY

[408] D. Moon. *The Grasslands of North America and Russia*, chapter 14, pp. 245–262. John Wiley & Sons, Blackwell Publishing, Hoboken NJ, 2012.

[409] C. M. Moore, M. M. Mills, K. R. Arrigo, et al. Processes and patterns of oceanic nutrient limitation. *Nature Geoscience*, 6:701–710, 2013.

[410] T. G. Morais, R. F. M. Teixeira, and T. Domingos. Detailed global modelling of soil organic carbon in cropland, grassland and forest soils. *PLoS One*, 14(e0222604), 2019.

[411] J. P. Morgan, L. H. Rupke, and W. M. White. The current energetics of Earth's interior: A gravitational energy perspective. *Frontiers in Earth Science*, 4, 2016.

[412] V. Morris and J. Jacquet. The animal agriculture industry, US universities, and the obstruction of climate understanding and policy. *Climatic Change*, 177(41), 2024.

[413] S. Mosier, S. Apfelbaum, P. Byck, et al. Adaptive multi-paddock grazing enhances soil carbon and nitrogen stocks and stabilization through mineral association in southeastern US grazing lands. *Journal of Environmental Management*, 288:112409, 2021.

[414] D. Mozaffarian. Dietary and policy priorities for cardiovascular disease, diabetes, and obesity. *Circulation*, 133(2):187–225, 2016.

[415] R. W. Mueller. *Solar Irradiance solar irradiance/irradiation, Global Distribution solar irradiance/irradiation global distribution*, pp. 553–583. Springer New York, 2013.

[416] A. A Musicus, D. D Wang, M. Janiszewski, et al. Health and environmental impacts of plant-rich dietary patterns: A US prospective cohort study. *Lancet Planetary Health*, 6:e892–e900, 2022.

[417] B. Mysen. Nitrogen in the Earth: Abundance and transport. *Progress in Earth and Planetary Science*, 6(38), 2019.

[418] H. N. Afrouzi, J. Ahmed, B. Mobin Siddique, et al. A comprehensive review on carbon footprint of regular diet and ways to improving lowered emissions. *Results in Engineering*, 18:101054, 2023.

[419] M. Nagao, H. Iso, K. Yamagishi, et al. Meat consumption in relation to mortality from cardiovascular disease among Japanese men and women. *European Journal of Clinical Nutrition*, 66:687–693, 2012.

[420] R. C. Nagy, E. J. Fusco, J. K. Balch, et al. A synthesis of the effects of cheatgrass invasion on US Great Basin carbon storage. *Journal of Applied Ecology*, 58:327–337, 2021.

[421] NatureServe. Biodiversity in focus: United States edition. Technical report, NatureServe, Arlington, VA, 2023.

[422] E. Neamatollahi, M. R. Jahansuz, D. Mazaheri, et al. *Intercropping*, pp. 119–142. Springer Netherlands, Dordrecht, 2013.

[423] A. G. Nelson, S. A. Quideau, B. Frick, et al. The soil microbial community and grain micronutrient concentration of historical and modern hard red spring wheat cultivars grown organically and conventionally in the black soil zone of the Canadian prairies. *Sustainability*, 3(3):500–517, 2011.

[424] M. Nestle. Paleolithic diets: A sceptical view. *Nutrition Bulletin*, 25(1):43–47, 2000.

[425] M. Nestle. *What to Eat*. North Point Press, 1st edition, 2007.

[426] M. Nestle. *Food Politics: How the Food Industry Influences Nutrition and H..* University of California Press, 2013.

[427] T. L. T. Nguyen, J. E. Hermansen, and L. Mogensen. Environmental consequences of different beef production systems in the EU. *Journal of Cleaner Production*, 18:756–766, 2010.

[428] R. W. Nicklas, I. S. Puchtel, R. D. Ash, et al. Secular mantle oxidation across the archean-proterozoic boundary: Evidence from v partitioning in komatiites and picrites. *Geochimica et Cosmochimica Acta*, 250:49–75, 2019.

[429] N. H. Niman. The carnivore's dilemma. *New York Times*, 10 2009.

[430] N. H. Niman. Defending 'foodies': A rancher takes a bite out of B. R. Myers. *The Atlantic*, 2, 2011.

[431] N. H. Niman. *Defending Beef: The Case for Sustainable Meat Production*. Chelsea Green Publishing, White River Junction VT, 2014.

[432] M. A. Nocco, R. A. Smail, and C. J. Kucharik. Observation of irrigation-induced climate change in the Midwest United States. *Global Change Biology*, 25(10):3472–3484, 2019.

[433] K. M. C. Nogoy, B. Sun, S. Shin, et al. Fatty acid composition of grain- and grass-fed beef and their nutritional value and health implication. *Food Science Animal Resource*, 42:18–33, 2022.

[434] M. Nordborg and E. Roos. Holistic management: A critical review of allan savory's grazing method. Technical report, SLU, Swedish University of Agricultural Sciences & Chalmers, Uppsala, Sweden, 2016.

[435] M. A. North, J. A Franke, B. Ouweneel, et al. Global risk of heat stress to cattle from climate change. *Environmental Research Letters*, 18(094027), 2023.

[436] J. Oben, E. Enonchong, S. Kothari, et al. Phellodendron and citrus extracts benefit joint health in osteoarthritis patients: A pilot, double-blind, placebo-controlled study. *Nutrition Journal*, 8, 2009.

BIBLIOGRAPHY 249

[437] B. Oberle, S. Bringezu, S. Hatfield-Dodds, et al. *Global Resources Outlook 2019: Natural Resources for the Future We Want*. The International Resource Panel, United Nations Environment Programme, Nairobi, Kenya, 2019.

[438] C. J. O'Bryan, A. R. Braczkowski, H. L. Beyer, et al. The contribution of predators and scavengers to human well-being. *Nature Ecology & Evolution*, 2:229–236, 2018.

[439] United States Department of Agriculture. Quick stats. https://quickstats.nass.usda.gov/, 2021.

[440] Department of Economic and Social Affairs Population Dynamics. World population prospects 2019, 2021. Online at https://population.un.org/wpp/Graphs/DemographicProfiles/Line/900; accessed 1 June 2021.

[441] Congresional Budget Office. Emissions of carbon dioxide in the transporation sector, 12 2022.

[442] G. Oliva, A. Cibils, P. Borrelli, et al. Stable states in relation to grazing in Patagonia: A 10-year experimental trial. *Journal of Arid Environments*, 40:113–131, 1998.

[443] G. Oliva, D. Ferrante, S. Puig, et al. Sustainable sheep management using continuous grazing and variable stocking rates in Patagonia: A case study. *Rangeland Journal*, 34:285–295, 2012.

[444] M. Oppenheimer, N. Oreskes, D. Jamieson, et al. University of Chicago Press, Chicago, 2019.

[445] N. Oreskes. *Why Trust Science?* Princeton University Press, Princeton, NJ, Woodstock, UK, Beijing, China, 2019.

[446] M. B. Osman, J. E. Tierney, J. Zhu, et al. Globally resolved surface temperatures since the last glacial maximum. *Nature*, 599:239–244, 2021.

[447] J. E. Overland. Causes of the record-breaking Pacific Northwest heatwave, late June 2021. *Atmosphere*, 12(1434), 2021.

[448] J. E. Overland, T. J. Ballinger, J. Cohen, et al. How do intermittency and simultaneous processes obfuscate the arctic influence on midlatitude winter extreme weather events? *Environmental Research Letters*, 16(043002), 3 2021.

[449] J. T. Overpeck and B. Udall. Climate change and the aridification of North America. *Proceedings of the National Academy of Sciences*, 117(22):11856–11858, 2020.

[450] P. Smith, M. Bustamante, H. Ahammad, et al. *Agriculture, Forestry and Other Land Use (AFOLU)*, chapter 11, pp. 811–922. IPCC-AR5. Cambridge University Press, Cambridge, UK and New York, NY, USA, 2014.

[451] I. Z. Palubski, A. L. Shields, R. Deitrick. Habitability and water loss limits on eccentric planets orbiting main-sequence stars. *Astrophysical Journal*, 890, 2020.

[452] A. Pan, Q. Sun, A. M. Bernstein, et al. Red meat consumption and mortality: Results from 2 prospective cohort studies. *Archives of Internal Medicine*, 172(7):555–563, 04 2012.

[453] A. Pan, Q. Sun, A. M. Bernstein, et al. Red meat consumption and risk of type 2 diabetes: 3 cohorts of US adults and an updated meta-analysis. *American Journal of Clinical Nutrition*, 94(4):1088–1096, 08 2011.

[454] P. K. Pandey and V. Pandey. Estimation of infiltration rate from readily available soil properties (RASPs) in fallow cultivated land. *Sustainable Water Resources Management*, 5:921–934, 2019.

[455] J. M. Paruelo, G. Pineiro, G. Baldi, et al. Carbon stocks and fluxes in rangelands of the Rio de la Plata basin. *Rangeland Ecology & Management*, 63:94–108, 2010.

[456] N. Pelletier, R. Pirog, and R. Rasmussen. Comparative life cycle environmental impacts of three beef production strategies in the upper Midwestern United States. *Agricultural Systems*, 103:380–389, 2010.

[457] W. Peng, E. M. Berry, and R. Goldsmith. Adherence to the Mediterranean diet was positively associated with micronutrient adequacy and negatively associated with dietary energy density among adolescents. *Journal of Human Nutrition and Dietetics*, 32:41–52, 2019.

[458] S. C. Penniston-Dorland, M. J. Kohn, and C. E. Manning. The global range of subduction zone thermal structures from exhumed blueschists and eclogites: Rocks are hotter than models. *Earth and Planetary Science Letters*, 428:243–254, 2015.

[459] B. L. Perryman, B. W. Schultz, and P. J. Meimanab. Forum: A change in the ecological understanding of rangelands in the Great Basin and Intermountain West and implications for management: Revisiting Mack and Thompson (1982). *Rangeland Ecology & Management*, 76:1–11, 2021.

[460] C. C. Peterson, K. A. Nagy, and J. Diamond. Sustained metabolic scope. *Proceedings of the National Academy of Sciences of the United States of America*, 87:2324–2328, 1990.

[461] R. T. Pierrehumbert and G. Eshel. Climate impact of beef: An analysis considering multiple time scales and production methods without use of global warming potentials. *Environmental Research Letters*, 10, 2015.

[462] R. T. Pierrehumbert. Short-lived climate pollution. *Annual Review of Earth and Planetary Sciences*, 42:341–379, 2014.

BIBLIOGRAPHY

[463] R. H. Pildes. The neglected value of effective government. *University of Chicago Legal Forum*, page 31, 2023. NYU School of Law, Public Law Research Paper No. 23–51.

[464] D. Pimentel and M. Pimentel. *Food, Energy, and Society*. University Press of Colorado, Niwot, CO, 1996.

[465] S. L. Pimm. *The Balance of Nature? Ecological Issues in the Conservation of Species and C.*. The University of Chicago Press, Chicago, 1992.

[466] G. Pineiro, J. M. Paruelo, M. Oesterheld, et al. Pathways of grazing effects on soil organic carbon and nitrogen. *Rangeland Ecology & Management*, 63:109–119, 2010.

[467] W. C. Pitman and M. Talwani. Sea-floor spreading in the North Atlantic. *GSA Bulletin*, 83:619–646, 1972.

[468] D. Plenet, J. Borg, C. Hilaire, et al. Agro-economic performance of peach orchards under low pesticide use and organic production in a cropping system experimental network in France. *European Journal of Agronomy*, 148:126866, 2023.

[469] M. Pollan. *The Omnivore's Dilemma*. Penguin Books, New York, 2006.

[470] M. Pollan. Big food strikes back: Why did the Obamas fail to take on corporate agriculture?, 10 2016.

[471] P. D. Polly. The politics of palaeontology: The creation, reduction, and restoration of Grand Staircase–Escalante and Bears Ears national monuments. *Geological Curator*, 11:436–454, 2022.

[472] L. Polsky and M. A. G. von Keyserlingk. Invited review: Effects of heat stress on dairy cattle welfare. *Journal of Dairy Science*, 100:8645–8657, 2017.

[473] L. C. Ponisio, L. K. M'Gonigle, K. C. Mace, et al. Diversification practices reduce organic to conventional yield gap. *Proceedings of the Royal Society B: Biological Sciences*, 282(1799):20141396, 2015.

[474] K. M. Pontoppidan, C. Salyk, A. Banzatti, et al. The nitrogen carrier in inner protoplanetary disks. *Astrophysical Journal*, 874, 2019.

[475] J. Poore and T. Nemecek. Reducing food's environmental impacts through producers and consumers. *Science*, 360:987–992, 2018.

[476] M. Prasad, S. Jayaraman, M. A. Eladl, et al. A comprehensive review on therapeutic perspectives of phytosterols in insulin resistance: A mechanistic approach. *Molecules*, 27(1595), 2022.

[477] F. Qian, M. C. Riddle, J. Wylie-Rosett, et al. Red and processed meats and health risks: How strong is the evidence? *Diabetes Care*, 43:265–271, 2020.

[478] A. Rahaman, A. Kumari, X.-A. Zeng, et al. The increasing hunger concern and current need in the development of sustainable food security in the developing countries. *Trends in Food Science and Technology*, 113:423–429, 2021.

[479] Jr. Ralph S. Paffenbarger and I-Min L. S. N. Blair. A history of physical activity, cardiovascular health and longevity: The scientific contributions of Jeremy N. Morris, DSC, DPH, FRCP. *International Journal of Epidemiology*, 30:1184–1192, 2001.

[480] C. Ratzke and J. Gore. Modifying and reacting to the environmental pH can drive bacterial interactions. *PLOS Biology*, 16(3):e2004248, 2018.

[481] B. M. Rau, D. W. Johnson, R. R. Blank, et al. Transition from sagebrush steppe to annual grass (bromus tectorum): Influence on belowground carbon and nitrogen. *Rangeland Ecology and Management*, 64:139–147, 2011.

[482] E. J. Raynor, J. D. Derner, D. J. Augustine, et al. Balancing ecosystem service outcomes at the ranch-scale in shortgrass steppe: The role of grazing management. *Rangelands*, 44(6):391–397, 2022.

[483] L. Reiley. Research group that discounted risks of red meat has ties to program partly backed by beef industry. *Washington Post*, 10, 2019. Accessed 24 June 2021.

[484] L. Reiley. Author of study saying red meat is fine failed to disclose industry funding, journal reveals. *Washington Post*, 1, 2020.

[485] M. Reiter. Observational constraints on the likelihood of 26Al in planet-forming environments. *Astonomy & Astrophysics*, 644(L1):1–4, 2020.

[486] S. Ren, C. Terrer, J. Li, et al. Historical impacts of grazing on carbon stocks and climate mitigation opportunities. *Nature Climate Change*, 14:380–386, 2024.

[487] G. Reynolds. Can athletes perform well on a vegan diet? *New York Times*, 6 2012.

[488] C. Richter. *The Sea of Grass*. Amereon, Mattituck, NY, 1936.

[489] A. Ricolleau, J.-P. Perrillat, G. Fiquet, et al. Phase relations and equation of state of a natural morb: Implications for the density profile of subducted oceanic crust in the Earth's lower mantle. *Journal of Geophysical Research: Solid Earth*, 115(B8):1–15, 2010.

[490] B. G. Ridoutt, G. Page, K. Opie, et al. Carbon, water and land use footprints of beef cattle production systems in Southern Australia. *Journal of Cleaner Production*, 73:24–30, 2014.

[491] D. L. Rife, J. O. Pinto, A. J. Monaghan, et al. NCAR global climate four-dimensional data assimilation (CFDDA) hourly 40 km reanalysis. 2014.

BIBLIOGRAPHY

[492] S. Rinella. *American Buffalo: In Search of a Lost Icon*. Spiegel and Grau, New York, 2009.

[493] S. Rinella. *Meat Eater: Adventures from the Life of an American Hunter*. Spiegel and Grau, New York, 2013.

[494] W. J. Ripple and R. L. Beschta. Trophic cascades in Yellowstone: The first 15 years after wolf reintroduction. *Biological Conservation*, 145:205–213, 2012.

[495] D. Rogerson. Vegan diets: Practical advice for athletes and exercisers. *Journal of the International Society of Sports Nutrition*, 14(36), 2017.

[496] T. Root and J. Schwartz. One thing you can do: Boil water efficiently. *The New York Times*, 5 2019.

[497] E. Ros, A. Singh, and J. H. O'Keefe. Nuts: Natural pleiotropic nutraceuticals. *Nutrients*, 13(9), 2021.

[498] C. A. Rotz, S. Asem-Hiablie, S. Place, et al. Environmental footprints of beef cattle production in the United States. *Agricultural Systems*, 169:1–13, 2019.

[499] J. E. Rowntree, R. Ryals, M. DeLonge, et al. Potential mitigation of midwest grass-finished beef production emissions with soil carbon sequestration in the United States of America. *Future of Food: Journal on Food, Agriculture and Society*, 4:S31–38, 2016.

[500] S. Roy and S. Bagchi. Large mammalian herbivores and the paradox of soil carbon in grazing ecosystems: Role of microbial decomposers and their enzymes. *Ecosystems*, 25:976–988, 2022.

[501] W. R. Teague, S. L. Dowhower, S. A. Baker, et al. Grazing management impacts on vegetation, soil biota and soil chemical, physical and hydrological properties in tall grass prairie. *Agriculture, Ecosystems & Environment*, 141:310–322, 2011.

[502] R. Rubin. Backlash over meat dietary recommendations raises questions about corporate ties to nutrition scientists. *Journal of the American Medical Association*, 323:401–404, 2020.

[503] W. F. Ruddiman. *Plows, Plagues, and Petroleum: How Humans Took Control of Climate*. Princeton University Press, Princeton, NJ, 2005.

[504] W. F. Ruddiman. The early anthropogenic hypothesis: Challenges and responses. *Reviews of Geophysics*, 45(4):RG4001, 2007.

[505] W. F. Ruddiman, D. Q. Fuller, J. E. Kutzbach, et al. Late Holocene climate: Natural or anthropogenic? *Reviews of Geophysics*, 54(1):93–118, 2016.

[506] W. F. Ruddiman, F. He, S. J. Vavrus, et al. The early anthropogenic hypothesis: A review. *Quaternary Science Reviews*, 240:106386, 2020.

[507] S. G. Ryan and A. J. Norton. *Stellar Evolution and Nucleosynthesis*. Cambridge University Press, Cambridge and New York, 2010.

[508] W. J. Sacks, B. I. Cook, N. Buenning, et al. Effects of global irrigation on the near-surface climate. *Climate Dynamics*, 33:159–175, 2009.

[509] M. S. V. Salles, S. C. Silva, L. C. Roma Junior, et al. Detection of heat produced during roughage digestion in ruminants by using infrared thermography. *Animal Production Science*, 50, 2017.

[510] A. Sanaei, E. J. Sayer, Z. Yuan, et al. Grazing intensity alters the plant diversity–ecosystem carbon storage relationship in rangelands across topographic and climatic gradients. *Functional Ecology*, 37:1–16, 2023.

[511] D. Sanders and T. van Erp. The physical demands and power profile of professional men's cycling races: An updated review. *International Journal of Sports Physiology and Performance*, 16(1):3–12, 2021.

[512] H. Sandoval-Insausti, Y.-H. Chiu, D. H. Lee, et al. Intake of fruits and vegetables by pesticide residue status in relation to cancer risk. *Environment International*, 156:106744, 2021.

[513] Y. Sano, N. Takahata, Y. Nishio, et al. Nitrogen recycling in subduction zones. *Geophysical Research Letters*, 25:2289–2292, 1998.

[514] A. Satija and F. B. Hu. Plant-based diets and cardiovascular health. *Trends in Cardiovascular Medicine*, 28:437–441, 2018.

[515] A. Savory. How to fight desertification and reverse climate change. 2013.

[516] A. Savory. Response to request for information on the "science" and "methodology" underpinning holistic management and holistic planned grazing. 2013.

[517] A. Savory and J. Butterfield. *Holistic Management: A New Framework for Decision Making*. Island Press, Washington, DC, 1998.

[518] J. D. Scasta, T. Gergeni, K. Maczko, et al. Adaptive grazing and animal density implications for stocking rate and drought in Northern mixed-grass prairie. *Livestock Science*, 269(105184), 2023.

[519] L. Schaefer and L. T. Elkins-Tanton. Magma oceans as a critical stage in the tectonic development of rocky planets. *Philosophical Transactions of the Royal Society A*, 376(37620180109), 2018.

[520] P. Schaff. *History of the Christian Church, V. II: Ante-Nicene Christianity. A.D. 100-325*. Independently published, 2017.

[521] A. Schersten, T. Elliott, C. Hawkesworth, et al. Hf-W evidence for rapid differentiation of iron meteorite parent bodies. *Earth and Planetary Science Letters*, 241(3):530–542, 2006.

[522] O. J. Schmitz, M. Sylven, T. B. Atwood, et al. Trophic rewilding can expand natural climate solutions. *Nature Climate Change*, 13, 2023.

[523] T. Schneider, P. A. O'Gorman, and X. J. Levine. Water vapor and the dynamics of climate changes. *Reviews of Geophysics*, 48, 2010.

[524] J. D. M. Schreel and K. Steppe. Foliar water uptake in trees: Negligible or necessary? *Trends in Plant Science*, 25(6):590–603, 2020.

[525] U. Schumann and H. Huntrieser. The global lightning-induced nitrogen oxides source. *Atmospheric Chemistry and Physics*, 7(14):3823–3907, 2007.

[526] H. P. Schwarcz and M. J. Schoeninger. *Stable Isotopes of Carbon and N. as Tracers for Paleo-Diet Reconstruction*, pp. 725–742. Springer Berlin Heidelberg, Berlin, Heidelberg, 2012.

[527] D. M. Schwindt, R. K. Bocinsky, S. G. Ortman, et al. The social consequences of climate change in the central Mesa Verde region. *American Antiquity*, 81:74–96, 2016.

[528] E. I. Scott, E. Toensmeier, F. Iutzi, et al. Policy pathways for perennial agriculture. *Frontiers in Sustainable Food Systems*, 6, 2022.

[529] J. W. Seaquist, E. Li Johansson, and K. A. Nicholas. Architecture of the global land acquisition system: Applying the tools of network science to identify key vulnerabilities. *Environmental Research Letters*, 9(114006), 2014.

[530] A. B. Searle and S. E. Meyer. Cattle trampling increases dormant season mortality of a globally endangered desert milkvetch. *Journal for Nature Conservation*, 56(125868), 2020.

[531] S. Secchi and M. Mcdonald. The state of water quality strategies in the Mississippi River basin: Is cooperative federalism working? *Science of The Total Environment*, 677:241–249, 2019.

[532] H. Seid and M. Rosenbaum. Low carbohydrate and low-fat diets: What we don't know and why we should know it. *Nutrients*, 11(11), 2019.

[533] T. Seki, T. Kamiya, K. Furukawa, et al. Nobiletin-rich Citrus reticulata peels, a kampo medicine for Alzheimer's disease: A case series. *Geriatrics & Gerontology International*, 13(1):236–238, 2013.

[534] M. J. Selinske, F. Fidler, A. Gordon, et al. We have a steak in it: Eliciting interventions to reduce beef consumption and its impact on biodiversity. *Conservation Letters*, 13(5):e12721, 2020.

[535] A. M. Senior, S. M. Solon-Biet, V. C. Cogger, et al. Dietary macronutrient content, age-specific mortality and lifespan. *Proceedings of the Royal Society B: Biological Sciences*, 286(1902):20190393, 2019.

[536] A. J. Severinsky and A. L. Sessoms. Methane versus carbon dioxide: Mitigation prospects. *International Journal of Environmental and Ecological Engineering*, 15:214–220, 2021.

[537] U. A. Shah and G. Merlo. Personal and planetary health—the connection with dietary choices. *JAMA, Journal of the American Medical Association*, 329:1823–1824, 2023.

[538] F. Shahidi, C. M. Liyana-Pathirana, and D. S. Wall. Antioxidant activity of white and black sesame seeds and their hull fractions. *Food Chemistry*, 99(3):478–483, 2006.

[539] A. Shepon, G. Eshel, E. Noor, et al. Energy and protein feed-to-food conversion efficiencies in the US and potential food security gains from dietary changes. *Environmental Research Letters*, 11(10):105002, 2016.

[540] G. J. Sheridan, H. B. So, R. J. Loch, et al. Use of laboratory-scale rill and interrill erodibility measurements for the prediction of hillslope-scale erosion on rehabilitated coal mine soils and overburdens. *Australian Journal of Soil Research*, 38:285–297, 2000.

[541] K. Sherren and C. Kent. Who's afraid of Allan Savory? Scientometric polarization on holistic management as competing understandings. *Renewable Agriculture and Food Systems*, 34:77–92, 2017.

[542] K. Sherren and C. Kent. Who's afraid of Allan Savory? Scientometric polarization on holistic management as competing understandings. *Renewable Agriculture and Food Systems*, 34:77–92, 2019.

[543] W. Shi, X. Huang, C. Mary Schooling, et al. Red meat consumption, cardiovascular diseases, and diabetes: A systematic review and meta-analysis. *European Heart Journal*, 44:2626–2635, 2023.

[544] V. S. Shilpa, R. Shams, K. K. Dash, et al. Phytochemical properties, extraction, and pharmacological benefits of naringin: A review. *Molecules*, 28(15), 2023.

[545] D. Shindell, L. Parsons, G. Faluvegi, et al. The important role of African emissions reductions in projected local rainfall changes. *NPJ Climate and Atmospheric Science*, 6(47), 2023.

[546] R. Shivakoti, M. L. Biggs, L. Djoussé, et al. Intake and sources of dietary fiber, inflammation, and cardiovascular disease in older US adults. *JAMA Network Open*, 5(3):e225012–e225012, 03 2022.

[547] A. Shropshire. *Grazing strategy effects on utilization, animal performance, aboveground production, species composition, and soil properties on Nebraska Sandhills Meadow* Master's thesis, University of Nebraska, Lincoln, 12, 2018.

BIBLIOGRAPHY

[548] E. R. Siirila-Woodburn, A. M. Rhoades, B. J. Hatchett, et al. A low-to-no snow future and its impacts on water resources in the Western United States. *Nature Reviews Earth & Environment*, 2:800–819, 2021.

[549] I. Simmonds and M. Li. Trends and variability in polar sea ice, global atmospheric circulations, and baroclinicity. *Annals of the New York Academy of Sciences*, 1504(1):167–186, 2021.

[550] J. M. Simonson, S. D. Birkel, K. A. Maasch, et al. Association between recent US northeast precipitation trends and Greenland blocking. *International Journal of Climatology*, 42:5682– 5693, 2022.

[551] P. A. D. Singer. *Animal Liberation*. Harper Collins, New York, 1975.

[552] R. Singh, S. Rastogi, and U. N. Dwivedi. Phenylpropanoid metabolism in ripening fruits. *Comprehensive Reviews in Food Science and Food Safety*, 9(4):398–416, 2010.

[553] P. W. Siri-Tarino, S. Chiu, N. Bergeron, et al. Saturated fats versus polyunsaturated fats versus carbohydrates for cardiovascular disease prevention and treatment. *Annual Review of Nutrition*, 35(1):517–543, 2015.

[554] B. A. Small, J. K. Frey, and C. C. Gard. Livestock grazing limits beaver restoration in Northern New Mexico. *Restoration Ecology*, 24:646–655, 2016.

[555] K. E. Smith, C. H. House, R. D. Arevalo, et al. Organometallic compounds as carriers of extraterrestrial cyanide in primitive meteorites. *Nature Communications*, 10(2777), 2019.

[556] L. C. Smith, G. M. MacDonald, A. A. Velichko, et al. Siberian peatlands a net carbon sink and global methane source since the early Holocene. *Science*, 303(5656):353–356, 2004.

[557] P. E. Smith, S. M. Waters, D. A. Kenny, et al. Effect of divergence in residual methane emissions on feed intake and efficiency, growth and carcass performance, and indices of rumen fermentation and methane emissions in finishing beef cattle. *Journal of Animal Science*, 99(11), 10 2021.

[558] P. Smith, K. Calvin, J. Nkem, et al. Which practices co-deliver food security, climate change mitigation and adaptation, and combat land degradation and desertification? *Global Change Biology*, 26:1532–1575, 2020.

[559] R. Smyth. Natureserve habitat models for imperiled species. 2021. Online at https://habitatsuitabilitymodeling-natureserve.hub.arcgis.com/pages/the-map -of-biodiversity-importance; accessed 10 June 2021.

[560] S. Sorooshian, J. Li, K. lin Hsu, et al. How significant is the impact of irrigation on the local hydroclimate in California's central valley? comparison of model results with ground and remote-sensing data. *Journal of Geophysical Research Atmospheres*, 116, 2011.

[561] J. F. Soussana, T. Tallec, and V. Blanfort. Mitigating the greenhouse gas balance of ruminant production systems through carbon sequestration in grasslands. *Animal*, 4:334–350, 2010.

[562] E. Sowula-Skrzynska, A. Borecka, J. Pawłowska, et al. Thermal stress influence on the productive and economic effectiveness of Holstein-friesian dairy cows in temperate climate. *Annals of Animal Science*, 23:887–896, 2023.

[563] Not specified. The Dublin Declaration of Scientists on the societal role of livestock. *Animal Frontiers*, 13:10, 4 2023.

[564] S. Spiegal, B. T. Bestelmeyer, D. W. Archer, et al. Evaluating strategies for sustainable intensification of US agriculture through the long-term agroecosystem research network. *Environmental Research Letters*, 13(034031), 2018.

[565] E. Spratt, J. Jordan, J. Winsten, et al. Accelerating regenerative grazing to tackle farm, environmental, and societal challenges in the upper Midwest. *Journal of Soil and Water Conservation*, 76:15A–23A, 2021.

[566] M. Springmann, M. Clark, D. Mason-D'Croz, et al. Options for keeping the food system within environmental limits. *Nature*, 562:519–525, 2018.

[567] J. E. Sprinkle, J. W. Holloway, B. G. Warrington, et al. Digesta kinetics, energy intake, grazing behavior, and body temperature of grazing beef cattle differing in adaptation to heat. *Journal of Animal Science*, 78:1608–1624, 2000.

[568] V. Stagno and Y. Fei. The redox boundaries of Earth's interior. *Elements*, 16:167–172, 2020.

[569] K. J. Stanienda-Pilecki. Crystals structures of carbonate phases with Mg in triassic rocks, mineral formation and transitions. *Scientific Reports*, 13(18759), 2023.

[570] P. Stanley, N. Sayre, and L. Huntsinger. Holistic management shifts ranchers' mental models for successful adaptive grazing. *Rangeland Ecology & Management*, 93:33–48, 2024.

[571] P. L. Stanley, J. E. Rowntree, D. K. Beede, et al. Impacts of soil carbon sequestration on life cycle greenhouse gas emissions in Midwestern USA beef finishing systems. *Agricultural Systems*, 162:249–258, 2018.

[572] P. Steinhardt. *The Second Kind of Impossible: The Extraordinary Quest for a New Form of Matter*. Simon & Schuster, New York, 2019.

[573] L. Stephens, D. Fuller, N. Boivin, et al. Archaeological assessment reveals Earth's early transformation through land use. *Science*, 365:897–902, 2019.

[574] B. D. Stocker, Z. Yu, C. Massa, et al. Holocene peatland and ice-core data constraints on the timing and magnitude of CO_2 emissions from past land use. *Proceedings of the National Academy of Sciences*, 114(7):1492–1497, 2017.

[575] T. F. Stocker, D. Qin, G.-K. Plattner, et al. *Climate Change 2013: The Physical Science Basis. Contribution of Working Group I to the Fifth Assessment Report of the Intergovernmental Panel on Climate Change*, page 1535. Cambridge University Press, Cambridge, and New York, 2014.

[576] H. Stommel. The westward intensification of wind-driven ocean currents. *Eos, Transactions of the American Geophysical Union*, 29:202–206, 1948.

[577] E. E. Stoody. Dietary guidelines for Americans 2020–2025. Technical report, United States Department of Agriculture, 2020.

[578] C. Streck, P. Keenlyside, and M. von Unger. The Paris agreement: A new beginning. *Journal for European Environmental & Planning Law*, 13(1):3–29, 2016.

[579] M. Sudermann, J. M. Galloway, D. R. Greenwood, et al. Palynostratigraphy of the lower paleogene Margaret formation at Stenkul Fiord, Ellesmere Island, Nunavut, Canada. *Palynology*, 45(3):459–476, 2021.

[580] S. Sumner, W. Wade, J. Selph, et al. Fertilization of established bahiagrass pasture in Florida. Technical Report 916, Florida Cooperative Extension Service Institute of Food and Agricultural Sciences University of Florida, Gainesville, 6, 1992.

[581] Y. Sun, J. Wang, S. Gu, et al. Simultaneous determination of flavonoids in different parts of Citrus reticulata 'chachi' fruit by high performance liquid chromatography—photodiode array detection. *Molecules*, 15(8):5378–5388, 2010.

[582] Z. Sun, P. Behrens, A. Tukker, et al. Shared and environmentally just responsibility for global biodiversity loss. *Ecological Economics*, 194(107339), 2022.

[583] J.-C. Svenning, R. Buitenwerf, and E. Le Roux. Trophic rewilding as a restoration approach under emerging novel biosphere conditions. *Current Biology*, 34(PR435-R451), 2024.

[584] M. Swain, L. Blomqvist, J. McNamara, et al. Reducing the environmental impact of global diets. *Science of The Total Environment*, 610-611:1207–1209, 2018.

[585] K. Z. Sweeney. *Prelude to the Dust Bowl: Drought in the Nineteenth-Century Southern Plains*. University of Oklahoma Press, Norman, 2016.

[586] T. Henriksen, A. Dahlback, S. H. Larsen, and J. Moan. Ultraviolet-radiation and skin cancer. Effect of an ozone layer depletion. *Photochemistry and Photobiology*, 51(5):579–582, 1990.

[587] S. Tao, R. M. Orellana Rivas, T. N. Marins, et al. Impact of heat stress on lactational performance of dairy cows. *Theriogenology*, 150:437–444, 2020.

[588] R. Teague. Deficiencies in the Briske et al. rebuttal of the Savory method. *Rangelands*, 36:37–38, 2014.

[589] R. Teague and U. Kreuter. Managing grazing to restore soil health, ecosystem function, and ecosystem services. *Frontiers in Sustainable Food Systems*, 29, 2020.

[590] R. Teague, F. Provenza, U. Kreuter, et al. Multi-paddock grazing on rangelands: Why the perceptual dichotomy between research results and rancher experience? *Journal of Environmental Management*, 128:699–717, 2013.

[591] W. R. Teague and S. I. Apfelbaum. *The Miracle of Grass*, pages 129–156. Springer Food and Health (FH) Series, Cham, 2023.

[592] M. M. Thiemens, P. Sprung, R. O. C. Fonseca, et al. Early moon formation inferred from hafnium–tungsten systematics. *Nature Geoscience*, 12:696–700, 2019.

[593] G. Thoma, B. Putman, M. Matlock, et al. Sustainability assessment of US beef production systems. Technical report, University of Arkansas Resiliency Center, 6, 2017.

[594] P. Thornton, G. Nelson, D. Mayberry, et al. Increases in extreme heat stress in domesticated livestock species during the twenty-first century. *Global Change Biology*, 27(22):5762–5772, 2021.

[595] G. Thunberg. *The Climate Book: The Facts and the Solutions*. Penguin Publishing Group, 2023.

[596] M. M. Tidball, K. G. Tidball, and P. Curtis. The absence of wild game and fish species from the usda national nutrient database for standard reference: Addressing information gaps in wild caught foods. *Ecology of Food and Nutrition*, 53:142–148, 2014.

[597] H. Tiessen. *Phosphorus in the Global Environment*, volume 7 of *Plant Ecophysiology*, chapter 1, pp. 1–7 Springer, Dordrecht, 2008.

[598] D. Tilman and M. Clark. Global diets link environmental sustainability and human health. *Nature*, 515:518–522, 2014.

[599] T. Australian Guide to Healthy Eating. Eat for health: Australian dietary guidelines. Technical report, Australia National Health and Medical Research Council, 2, 2013.

[600] J. E. Todd, E. Leibtag, and C. Penberthy. Geographic differences in the relative price of healthy foods. Technical Report EIB-78, United States Department of Agriculture, Economic Research Service, 6, 2011.

[601] A. Tomaino, M. Martorana, T. Arcoraci, et al. Antioxidant activity and phenolic profile of pistachio (pistacia vera l., variety bronte) seeds and skins. *Biochimie*, 92(9):1115–1122, 2010. Advances in Biomolecular and Medicinal Chemistry.

[602] F. Tong, P. Jaramillo, and I. M. L. Azevedo. Comparison of life cycle greenhouse gases from natural gas pathways for medium and heavy-duty vehicles. *Environmental Science & Technology*, 49(12):7123–7133, 2015.

BIBLIOGRAPHY   261

[603] C. C. Treat, M. C. Jones, L. Brosius, et al. The role of wetland expansion and successional processes in methane emissions from northern wetlands during the holocene. *Quaternary Science Reviews*, 257(106864), 2021.

[604] E. Trinkaus. Late pleistocene adult mortality patterns and modern human establishment. *Proceedings of the National Academy of Sciences*, 108(4):1267–1271, 2011.

[605] F. N. Tubiello, C. Rosenzweig, G. Conchedda, et al. Greenhouse gas emissions from food systems: Building the evidence base. *Environmental Research Letters*, 16, 2021.

[606] F. Turati, F. Concina, M. Rossi, et al. Association of prebiotic fiber intake with colorectal cancer risk: The prebiotica study. *European Journal of Nutrition*, 62:455–464, 2023.

[607] H. Twinomuhwezi, A. C. Godswill, and D. Kahunde. Extraction and characterization of pectin from orange (Citrus sinensis), lemon (Citrus limon) and tangerine (Citrus tangerina). *American Journal of Physical Sciences*, 1(1):17–30, March 2020.

[608] M. Umair, D. Kim, R. L. Ray, et al. Evaluation of atmospheric and terrestrial effects in the carbon cycle for forest and grassland ecosystems using a remote sensing and modeling approach. *Agricultural and Forest Meteorology*, 295(108187), 2020.

[609] S. Urbanski, C. Barford, S. Wofsy, et al. Factors controlling CO_2 exchange on timescales from hourly to decadal at Harvard Forest. *Journal of Geophysical Research: Biogeosciences*, 112(G2):1–25, 2007.

[610] Climate indicators explorer, atmospheric concentrations of greenhouse gases. 2023.

[611] USDA. Pesticide data program service annual summary, calendar year 2021. Technical report, United States Department of Agriculture, Agricultural Marketing Service, 12 2022.

[612] USDA-NASS. Meat animals production, disposition, and income 2020 summary. report, United States Department of Agriculture, National Agricultural Statistics Service, Washington, DC, 4, 2021.

[613] G. K. Vallis. *Atmospheric and O. Fluid Dynamics: Fundamentals and L.-scale Circulation*. Cambridge University Press, Cambridge, 1st edition, 2006.

[614] H. Valls-Fox, S. Chamaille-Jammes, M. de Garine-Wichatitsky, et al. Water and cattle shape habitat selection by wild herbivores at the edge of a protected area. *Animal Conservation*, 21(5):365–375, 2018.

[615] K. C. H. van Ginkel, W. J. W. Botzen, M. Haasnoot, et al. Climate change induced socio-economic tipping points: Review and stakeholder consultation

for policy relevant research. *Environmental Research Letters*, 15(2):023001, 1 2020.

[616] E. J. Van Loo, V. Caputo, and J. L. Lusk. Consumer preferences for farm-raised meat, lab-grown meat, and plant-based meat alternatives: Does information or brand matter? *Food Policy*, 95:101931, 2020.

[617] S. van Vliet, F. D. Provenza, and S. L. Kronberg. Health-promoting phytonutrients are higher in grass-fed meat and milk. *Frontiers in Sustainable Food Systems*, 4, 2021.

[618] G. Vandenbroucke. The US westward expansion. *International Economic Review*, 49:81–110, 2008.

[619] J. Venturini, M. P. Ronco, and O. M. Guilera. Setting the stage: Planet formation and volatile delivery. *Space Science Reviews*, 216(86), 2020.

[620] F. Vernia, S. Longo, G. Stefanelli, et al. Dietary factors modulating colorectal carcinogenesis. *Nutrients*, 13(143), 2021.

[621] H. Vollstaedt, K. Mezger, and Y. Alibert. Carbonaceous chondrites and the condensation of elements from the Solar Nebula. *Astrophysical Journal*, 897, 2020.

[622] R. Volpe, E. Roeger, and E. Leibtag. How transportation costs affect fresh fruit and vegetable prices. Technical Report Economic Research Report Number 160, U.S. Department of Agriculture Forest Service, Economic Research Service, Washington DC, 11 2013.

[623] M. W. Broadley, D. V. Bekaert, L. Piani, et al. Origin of life-forming volatile elements in the inner solar system. *Nature*, 611(7935):245–255, 2022.

[624] D. D. Wang, Y. Li, S. N. Bhupathiraju, et al. Fruit and vegetable intake and mortality. *Circulation*, 143(17):1642–1654, 2021.

[625] H. S. Wang, C. Lineweaver, and T. R. Ireland. The elemental abundances (with uncertainties) of the most earth-like planet. *Icarus*, 299:460–474, 2018.

[626] M. Wang, M. Wander, S. Mueller, et al. Evaluation of survey and remote sensing data products used to estimate land use change in the United States: Evolving issues and emerging opportunities. *Environmental Science and Policy*, 129:68–78, 2022.

[627] P. Wang. Low-latitude forcing: A new insight into paleo-climate changes. *The Innovation*, 2(100145), 2021.

[628] X. Wang, X. Lin, Y. Y Ouyang, et al. Red and processed meat consumption and mortality: Dose–response meta-analysis of prospective cohort studies. *Public Health Nutrition*, 19:893–905, 2015.

[629] Y. Wang, S. Yang, A. Sanchez-Lorenzo, et al. A revisit of direct and diffuse solar radiation in China based on homogeneous surface observations: Climatology,

trends, and their probable causes. *Journal of Geophysical Research: Atmospheres,* 125(e2020JD032634), 2020.

[630] Y. Wang, G. Zhou, and B. Jia. Modeling SOC and NPP responses of meadow steppe to different grazing intensities in Northeast China. *Ecological Modelling,* 217:72–78, 2008.

[631] S. G. Wannamethee, B. J. Jefferis, L. Lennon, et al. Serum conjugated linoleic acid and risk of incident heart failure in older men: The British regional heart study. *Journal of the American Heart Association,* 7(e006653), 2018.

[632] website staff. Climateone, 6 2021. Online at https://www.climateone.org /people/nicolette-hahn-niman; accessed 24 June 2021.

[633] C. L. Weber and H. S. Matthews. Food-miles and the relative climate impacts of food choices in the United States. *Environmental Science & Technology,* 42:3508–3513, 2008.

[634] K. T. Weber and B. S. Gokhale. Effect of grazing on soil-water content in semi-arid rangelands of southeast Idaho. *Journal of Arid Environments,* 75:464–470, 2011.

[635] J. Weim, P. A. Dirmeyer, D. Wisser, et al. Where does the irrigation water go? An estimate of the contribution of irrigation to precipitation using merra. *Journal of Hydrometeorology,* 14:275–289, 2013.

[636] R. F. Weingroff. Moving the goods: As the interstate era begins. Technical report, U.S. Department of Transportation, Federal Highway Administration, 1200 New Jersey Avenue SE, Washington DC 20590, 2017.

[637] N. Weitzel, H. Andres, J.-P. Baudouin, et al. Towards spatio-temporal comparison of simulated and reconstructed sea surface temperatures for the last deglaciation. *Climate of the Past,* 20:865–890, 2024.

[638] F. H. Westheimer. Why nature chose phosphates. *Science,* 235:1173–1178, 1987.

[639] M. Wild, D. Folini, M. Z. Hakuba, et al. The energy balance over land and oceans: An assessment based on direct observations and cmip5 climate models. *Climate Dynamics,* 44:3393–3429, 2015.

[640] G. Wiles, H. Allen, and G. Hayes. Wolf conservation and management plan. 12, 2011.

[641] J. K. Willenbring, A. T. Codilean, K. L. Ferrier, et al. Short communication: Earth is (mostly)flat, but mountains dominate global denudation: Apportionment of the continental mass flux over millennial timescales, revisited. *Earth Surface Dynamics Discussions,* 2:1–17, 2014.

[642] W. Willett, J. Rockström, B. Loken, et al. Food in the Anthropocene: The eat–lancet commission on healthy diets from sustainable food systems. *Lancet,* 393:447–492, 2019.

[643] W. Willett and P. J. Skerrett. *Eat, Drink and Be Healthy: The Harvard Medical School Guide to Healthy Eating*. Free Press, New York and London, 2017.

[644] A. Park Williams, B. I. Cook, and J. E. Smerdon. Rapid intensification of the emerging southwestern North American megadrought in 2020–2021. *Nature Climate Change*, 12(3):232–234, 2022.

[645] A. Park Williams, B. I. Cook, J. E. Smerdon, et al. The 2016 southeastern US drought: An extreme departure from centennial wetting and cooling. *Journal of Geophysical Research Atmospheres*, 122:10,888–10,905, 2017.

[646] A. C. Williams and L. J. Hill. Meat and nicotinamide: A causal role in human evolution, history, and demographics. *International Journal of Tryptophan Research*, 10(1178646917704661), 2017.

[647] E. O. Wilson. *Biophilia*. Harvard University Press, Cambridge, MA, and London, 1986.

[648] E. O. Wilson. A personal brief for the wildlands project. In Tom Butler and Reed Noss, editors, *The Wildlands Project*, volume 10 of *Wild Earth Society*, chapter 0, page 114. Dave Foreman, Richmond VT, 1st edition, 2000.

[649] K. C. Wirnitzer. Vegan diet in sports and exercise–Health benefits and advantages to athletes and physically active people: A narrative review. *International Journal of Open Access Sports and Exercise Medicine*, 6(165), 2020.

[650] A. Wolf and R. M. Mitchell. Leveraging historic cattle exclosures to detect evidence of state change in an arid rangeland. *Rangeland Ecology & Management*, 78:26–35, 2021.

[651] A. Wolk. Potential health hazards of eating red meat. *Journal of Internal Medicine*, 281:106–122, 2017.

[652] B. J. Wood, D. J. Smythe, and T. Harrison. The condensation temperatures of the elements: A reappraisal. *American Mineralogist*, 104:844–856, 2019.

[653] R. Wordsworth and L. Kreidberg. Atmospheres of rocky exoplanets. *Annual Review of Astronomy and Astrophysics*, 60:159–201, 2022.

[654] D. Worster. *Dust Bowl: The Southern Plains in the 1930s*. Oxford University Press, Oxford and New York, 25th anniversary edition, 2004.

[655] A. M. L. Würtz, M. U. Jakobsen, M. L. Bertoia, et al. Replacing the consumption of red meat with other major dietary protein sources and risk of type 2 diabetes mellitus: A prospective cohort study. *American Journal of Clinical Nutrition*, 113(3):612–621, 10 2021.

[656] X. Xu, J. Zhang, Y. Zhang, et al. Associations between dietary fiber intake and mortality from all causes, cardiovascular disease and cancer: A prospective study. *Journal of Transllational Medicine*, 20(344), 2022.

BIBLIOGRAPHY

[657] B. Yablonski. Bisonomics. https://www.perc.org/perc_reports/volume-25-no-3 -fall-2007/, 9 2007. PERC Reports, volume 25, No. 3.

[658] S. Yamada, M. Shirai, K. Ono, et al. Beneficial effects of a nobiletin-rich formulated supplement of sikwasa (c. depressa) peel on cognitive function in elderly Japanese subjects; a multicenter, randomized, double-blind, placebo-controlled study. *Food Science & Nutrition*, 9(12):6844–6853, 2021.

[659] K. Yamagishi, K. Maruyama, A. Ikeda, et al. Dietary fiber intake and risk of incident disabling dementia: The circulatory risk in communities study. *Nutritional Neuroscience*, 26(2):148–155, 2023.

[660] R. Yan and J. C. Yu Louie. Paleolithic diet as a potential dietary management option for type 2 diabetes: A scoping review. *Human Nutrition & Metabolism*, 36(200264), 2024.

[661] Y. Yang, D. Tilman, G. Furey, et al. Soil carbon sequestration accelerated by restoration of grassland biodiversity. *Nature Communications*, 10(718), 2019.

[662] Y. Yang, C. Liu, N. Ou, et al. Moisture transport and contribution to the continental precipitation. *Atmosphere*, 13(1694), 2022.

[663] Y. Yang and M. L. Roderick. Radiation, surface temperature and evaporation over wet surfaces. *Quarterly Journal of the Royal Meteorological Society*, 145:1118–1129, 2019.

[664] V. Yee. Tunisian cave village empties out in face of drought and modernity's draw, *The New York Times*, 1 2023.

[665] L.-Kin Yeung, D. M. Alschuler, M. Wall, et al. Multivitamin supplementation improves memory in older adults: A randomized clinical trial. *The American Journal of Clinical Nutrition*, 118(1):273–282, 2023.

[666] X. Yu, C. Keitel, Y. Zhang, et al. Global meta-analysis of nitrogen fertilizer use efficiency in rice, wheat and maize. *Agriculture, Ecosystems & Environment*, 338(108089), 2022.

[667] Y. Yu, R. Li, L. Pu, et al. Citrus tangerine pith extract alleviates hypoxia-induced ileum damage in mice by modulating intestinal microbiota. *Food Function*, 14:6062–6072, 2023.

[668] J. G. Zaller, A. Oswald, M. Wildenberg, et al. Potential to reduce pesticides in intensive apple production through management practices could be challenged by climatic extremes. *Science of The Total Environment*, 872(162237), 2023.

[669] S. Zelber-Sagi, D. Ivancovsky-Wajcman, N. F. Isakov, et al. High red and processed meat consumption is associated with non-alcoholic fatty liver disease and insulin resistance. *Journal of Hepatology*, 68(6):1239–1246, 2018.

[670] D. Zeraatkar, B. C. Johnston, J. Bartoszko, et al. Effect of lower versus higher red meat intake on cardiometabolic and cancer outcomes. *Annals of Internal Medicine*, 171(10):721–731, 2019.

[671] A. L. Zerkle and S. Mikhail. The geobiological nitrogen cycle: From microbes to the mantle. *Geobiology*, 15:343–352, 2017.

[672] T. Zhan, H. Zhao, J. Zhang, et al. Differential effects of grazing intensity on carbon sequestration in arid versus humid grasslands across China. *Science of the Total Environment*, 881(163221), 2023.

[673] L. Zhang, B. Wylie, L. Ji, et al. Carbon dynamics (2000–2006) over the northern Great Plains grasslands. In *International Grassland Congress Proceedings, XXI international grassland congress / VIII international rangeland congress, Hohhot, China from June 29 through July 5, 2008*, page 913, Guangzhou, Guangdong, China, 2008. Guangdong People's Publishing House.

[674] S. Zhang, G. Huang, and Y. Zhang. Sustained productivity and agronomic potential of perennial rice. *Nature Sustainability*, 6:28–38, 2023.

[675] X. Zhang, B. B. Ward, and D. M. Sigman. Global nitrogen cycle: Critical enzymes, organisms, and processes for nitrogen budgets and dynamics. *Chemical Reviews*, 120:5308–5351, 2020.

[676] Z. Zhang, J. Gong, B. Wang, et al. Regrowth strategies of Leymus chinensis in response to different grazing intensities. *Ecological Applications*, 30(e02113), 2020.

[677] M. Zhou, P. W. G. Groot Koerkamp, T. T. T. Huynh, et al. Evaporative water loss from dairy cows in climate-controlled respiration chambers. *Journal of Dairy Science*, 106:2035–2043, 2023.

[678] M. Zhou, P. W. G. Groot Koerkamp, T. T. T. Huynh, et al. Development and evaluation of a thermoregulatory model for predicting thermal responses of dairy cows. *Biosystems Engineering*, 223, 295–308, 2022.

[679] P. Zhu, J. Burney, J. Chang, et al. Warming reduces global agricultural production by decreasing cropping frequency and yields. *Nature Climate Change*, 12:1016–1023, 2022.

[680] K. D. Zink and D. E. Lieberman. Impact of meat and lower palaeolithic food processing techniques on chewing in humans. *Nature*, 531:500–503, 2016.

[681] C. Zugravu, A. Macri, N. Belc, et al. Efficacy of supplementation with methylcobalamin and cyancobalamin in maintaining the level of serum holotranscobalamin in a group of plant based diet (vegan) adults. *Experimental and Therapeutic Medicine*, 22(993), 2021.

Index

Adaptive multi-paddock grazing, 51, 87, 91, 93–97. *See also* Cattle grazing
Agribusiness, 212–215
  food system, impact on, 214–215
  modern societies and, 213
  policies, principles of, 212
  processing and nutritional degradation, 212–213
Agricultural nutrient rich runoff, 12
Agricultural pollution, 166–168
  biogeochemical cycle and, 149
  in Chesapeake Bay, 149
Agricultural regionalization, 146–149
  buying local, 146–147
  emission-related benefit of, 147, 148
  local food, favoring, 147–148
  transportation, reduce, 147
Agricultural shifts, 145–146
Agriculture
  climate change and, 148–149, 163–166
  denitrification and nitrogen loss in, 166–167
  Earth's natural cycles in, role of, 172–173

evaporative cooling as energy flux in, 185
  impact on Earth's cycles, 189–192
  interconnected water and nutrient systems in, 160–161
  orbital variations and their effects on, 154–155
  regional specificity in, 146–147
  soil and water dynamics in, 161–164
  as system of fluxes and reservoirs, 171–172
Agriculture, impact on environment, 11–12
  beef *vs.* poultry, 12–15
  disinformation and public perception of, 120
  food-related challenges and opportunities, 11–12
  global biodiversity loss, driver of, 12
  industrialized global, 14
  meat with plant alternatives, 15
  planetary impacts of, 4
  rangeland, use of, 16–22
  soil erosion, 11–12

INDEX

Agriculture, planetary factors and,
151–153
  climate cycles, 154
  Earth's physical characteristics, role of,
152–153
  food's planetary impact, 151–152
  glacial cycles and soil formation,
154–155
  plate tectonics on agricultural land
distribution, effect of, 153
Air specific humidity, climatological,
125
Almond protein, 42
Alternative protein sources, 82–85
  adoption of, national-scale impact of,
83
  animal-based protein alternatives, 82
  plant-based protein adoption, 84
  plant-based protein alternatives, 81
  production of, environmental cost of,
85
American bison, 56–67
Animal protein, 33, 41, 42, 214. *See also*
Beef
Apple farming
  subsidence, effects of, 143–144
  Washington, financial advantages of,
145
  westerlies and the Cascades in, role of,
143
Apple production, 141–146
  distribution of, 141
  geographical and climatic conditions,
impact of, 142–145
  geographic distribution in the US,
141–142
  market forces shaping apple-growing
regions, 142
  of New York, 141, 145
  prediction of, 145
  subsidence and, 143–144
  in Washington, 141, 142–145
Apple protein, 42

Aryal, D. R., 54
Atmospheric and Oceanic Fluid Dynamics
(G. Vallis), 138. *See also* Vallis, Geoff
Atmospheric circulation, 175
Atmospheric convergence over land,
175–176
Atmospheric divergence over oceans,
175–176

B_{12}, 214
  insufficiency, 44
  intake of, vegans and, 34
  nutritional yeast source of, 201
Beef, 41–71
  almond *vs.*, 42
  alternatives of, 43, 77–78
  apple *vs.*, 42
  comparative footprint, 41–43
  contributions global warming, 14
  environmental burdens of diet, 43
  environmental costs of, 44–45
  forgo beef, 22
  forgone, 36, 82
  grass-finished *vs.* conventional, 79
  grassland management model, 51–56
  greenhouse gas emissions, mitigation
of, 67
  nutritional value of, 33–35
  poultry *vs.*, 12–15
  value developed, confusion, 23–30
Beef consumption, 32
  cattle feed rations, 46
  and degenerative disease prevalence,
28, 29
  human evolution and, 35–38
  maintaining, 73
  methane and carbon dioxide
emission, 59–60, 63, 64
  Niman assertion, 28–29
  reduction in, biodiversity benefits of,
83
  rising, cattle inventory and, 28
Beef production, 74–85

environmental cost of, 85
land use and, 80
net carbon costs of, 79
reduction of, impact of, 85
Beef protein, 14
average consumption by Americans, 73
carbon dioxide emission, 52, 55–56, 68
production, 18, 41–42
Beef-to-poultry shift, 13–14
Berkshire Food CoOp, 199
Big Bang birth of universe, 103
Biodiversity, 200
Biogeochemical cycle, 148, 177–179
agricultural pollution and, 149
energy consumption, 173
material exchanges, 176
Biomass removal, as accelerated natural flux, 193
Bittman, Mark, 214
Bonny Doon Ecological Reserve, 133
Bos Indicus, 123
Bos Taurus, 123
Brewton, Jim (fictional character in *Sea of Grass*), 26
Briske, D. D., 89, 90
Broecker, Wally, 4, 5

Carbon
chemical flexibility of, 102–103
nucleosynthesis of, 105
oxidation states of, 103
Carbon-12, 109
Carbon balance, 25
beef alternatives based on, 55
grazing cessation on, impact of, 79
Carbon cycle
biochemical complexity of, 103
biological *vs.* geological components, 176–177
climate change and, 193–196
energy sources, 176
geological part of, 114

oceanic CO_2 solubility, role of, 176–177
Carbon dioxide emission
beef consumption, 59–60, 63, 64
beef emissions *vs.* alternative proteins, 82
beef protein, 52, 55–56, 68
land rewilding and reforestation impact on, 81
reduction through dietary shifts, 83
Carbon emissions savings, from forgoing beef, 81
Carbon neutrality, 64, 66
Carbon sequestration
cattle grazing, 53, 80
in grazed lands, 87
rangeland, 47
reforestation and, 78–79
in semi-arid grasslands and shortgrass steppes, 81
Carbon storage
dryland soil, 47
forests *vs.* grasslands, 78
soil. *See* Soil carbon content
topsoil, 97
Carbon uptake
atmosphere-to-soil, 50
for century-old forest, 78
conservative cropland reforestation, 80, 83
by grazelands, 67, 80
grazing management, 50–51
Harvard Forest, 78
long-term, 78
production in the United States, 41
rangeland ecosystems, 71
reforestation-related, 83
Cascades (Mountain Range), 143–144
Cattle grazing, 18–22. *See also* Ruminant grazing
along Gila and San Francisco Rivers, 20
argument promoting, 46–56
atmospheric carbon sink, 70

Cattle grazing (cont.)
  belowground carbon sequestration, promote, 53
  carbon sequestration, enhance, 51, 64–65, 66
  carbon uptake by rangeland ecosystems, reduces, 71
  rangelands, carbon store in, 50
  soil organic carbon, impact on, 48–49
  woody vegetation sequestration and, 80
Cattle heat regulation
  physics of, 130–133
  energy consuming activity, 136
  evaporative cooling, 123–124
  feed energy and, 134–135
  grass eating, impact of, 135–136
  human *vs.* cattle heat dissipation, 122–123
  lactating cows, metabolic heat generation in, 122
  relative humidity in, role of, 139
  roughage consumption on, impact of, 136
  sweating and, 123–124, 132–133
  wind system on, impact of, 140
Chesapeake Bay
  anoxia, contribution to, 148
  heavily nutrient polluted, 149
CLA. *See* Linoleic fatty acid
Clausius–Clapeyron relation, 127
Cooking, 208–210
  gas *vs.* electric cooking, 209
  kimchi, 210
  lentils and quinoa, 208–210
  microwave efficiency, 209–210
  salad preparation, 210–212
Core
  Earth's, 109, 112–113
  of older stars, 103
  stellar, 107, 108–110
Corn, 152–160. *See also* Agriculture

biological productivity of, nutrients required for, 158–159
dominance in U.S. agricultural exports, 152–153
as dominant crop, 152
Earth eccentricity and, 153–154
Earth's rotation rate, impact of, 153
environmental requirements, 156–160
growth of, role of carbon dioxide in, 156
industrial agriculture, role in, 152–153
mineral soil formation, 154
nutrient requirements for, 158–160
productivity, 156
thick soil required for, 153
water requirements of, 157–158
water retention, impact of gravitational field on, 152–153
yields, 155–156
Corn farming. *See also* Agriculture
  land allocation for, 152
  water demand for, 158
Croplands, 18, 41, 42
  allocated to corn, 152
  carbon favorability of reforesting, 80–81
  reforestation of, 78, 80–81, 82
Crop productivity, environmental factors influencing, 163

Dairy farming
  atmospheric subsidence, impact of, 121–130
  climate and environmental influences on, 137–141
  economic impact of heat stress on, 128–129
  geographic trends in, 120–122
  heat stress in dairy cattle, 122–124
Dairy industry migration, 120–133
  dairy production in New England, decline of, 121

evaporation and, 138–140
growth of dairy farming in Texas, 121
shift from lush to arid lands, 133,
140–141
Dairy productivity
heat production and, 136
in humid *vs.* arid states, 140–141
sustainable, role of subsidence in, 141
Dar, Shimon, 5
Dead zone, 191–192
Defending Beef, 38
Deforestation, 3
Department of the Interior's Bureau of
Land Management, 20
Dietary choices, 3, 22, 148
breakfast choices, 201, 210
environmental impact, 3, 23, 185,
199–201
greenhouse gas emissions, 199–200
of organic *vs.* conventional foods,
205–208
thoughtful, benefits of, 200
Dobzhansky, T., 140
Dust Bowl, 70

Earth's composition, 110
Earth's internal energy, 180–181
Earth's natural cycles
agriculture, role in, 172–173
energy sources driving, 173–175
Earth system, 3, 181
agriculture, impact of, 190
complexity of, 120
ocean–atmosphere carbon fluxes,
197
Eating, 151
Elements
abundance modified by planetary
processes, 110–113
atomic mass dependence 7–40, 103
life-essential. *See* Life-essential
elements

planetary formation of, 108–111
El Niño, 184
Essential nutrients for agriculture, 101

Feed energy efficiency in cattle,
134–135
Fennel, 201
Fertilizers
agricultural discharge, 148, 149
agricultural productivity and, 102
nitrogenous, 42
water pollution from, 166
Food choices. *See* Dietary choices
Food–diet–agriculture–environment,
119–120
Food industrial processing, 204–205
Food miles, 151
Food production, 21–22, 160, 189, 207.
See also Food system
carbon emissions from, 148
cattle grazing and, 92
croplands, 17, 52
in developing countries, 12
human, 45, 52, 93, 96
quantifying, 69
rangelands, 17
Food-related challenges and
opportunities, 11–12
Food sourcing, environmental impact
of, 148–149
Food system. *See also* Food production
American, 201
current, failure of, 212
personal choice and, 213–214
redesigned alternative, 46
technologies and, 141–142
Food waste reduction, 201–203
forgo beef, 22
Forgone beef, 36, 82
Fossil fuel
combustion, 3
energy, herd consumed of, 18

Gardner, C. D., 34
Gila Box Riparian National
    Conservation Area, 20
Glacial cycles, 2
Glaciation events, 1–2
Global warming. *See also* Greenhouse
    gas emissions
  anthropogenic, 1
  beef contributions, 14
Grass consumption
  hindgut *vs.* foregut fermentation, 136
  and nutritional challenges, 136
Grass fed beef, 68–69
  as net carbon source, 85
Grass-finished beef, 74
  carbon dioxide emission, 55
  conventional beef *vs.*, 79
  cropland rewilding *vs.*, 80
  eater, 77
  operations as carbon sources, 81
  replaced with alternatives, 83
Gravitational field, 152–153
Grazing cattle
  beneficial interactions with grassland,
    87
  feed aerobic oxidation, derive energy
    from, 136
  mob, 98
  nutritional and environmental
    acceptability of, 23–24
  rangeland sequestration by, 80–81
  trampling and manure deposition by,
    53
Great Oxidation Event, 7
Great Plains, 88, 137
  annual mean temperature 3 km above,
    139
  soil minerals of last glacial period, 155
  subsidence over, 138
Greenhouse gas emissions, 2, 70
  beef cattle, mitigation of, 67
  climate change and, 187
  current food system and, 212

food production and, 21–22
net global, feeding and, 11
reducing, 199–200
waste management and, 201

Helium-4, 104, 108
  binding energy of, 106, *107*
  nucleus, fusion of protons into, 105,
    106
  stable, 107
Holistic Planned Grazing, 89
Holocene, 3
  high interglacial warmth during, 2
  interglacial, 1
  methane and carbon dioxide records,
    6, 7
Hydrogen fusion, 108–109
Hydrological cycle
  energy source for, 173
  hijacking of, 162
  solar energy in, role of, 173–174
Hyper-processed food, 34

Ice age(s), 1
  mineral-rich soil, impact on, 154–155
  role in soil development, 155
Industrial Revolution, 3
Infrared radiation exchanges, 184–185
Interglacials., 2, 3
Israel, 4

Katz, D. L., 34

Last Glacial Maximum, 1, 2
Lentils and quinoa cooking, 208–210
Life-essential elements
  chemical flexibility of, 102–103
  cosmochemical background and
    element formation, 104–107
  reactivity of, 102
  in stars, 104
Linoleic fatty acid (CLA), 33
Livestock grazing. *See* Cattle grazing

INDEX

273

Lush grasslands-based beef operation
  alternative options of, 77–85
  beef production, 74
  carbon sequestration, 75, 76–77
  direct operational emissions during
    production, 75
  land needs of, 75–76
  resource needs of grass-finished beef
    eater, 77

Macronutrients balance, 203–204
Magma Ocean, 112
Micronutrients, 33, 202–204
  health promoting, 203
  ideal ratio, 203
  in Pleistocene fauna, 37
  USDA, 205
Middle East, 4–6
Mid-ocean ridges, 178
Milo, Ron, 12
Mob grazing, 98
Momentum flux, from atmosphere to
  land and ocean, 183–184
Mosier, S., 93
Multi-paddock grazing, adaptive, 51, 87,
  91, 93
Murphy, Tom, 208–209

Nebraska Sandhills meadow, 98
Nemecek, T., 68, 69, 74
Niman, Nicolette Hahn, 27–29
  Defending Beef, 27
  Diet for a Small Planet, 27
Nitrogen
  biological functions, role in, 101
  chemical flexibility of, 102
  dominance, 111
  galactic and solar system, 109
  in mature Earth, 114–117
  nucleosynthesis of, 105
  in ocean floor sediments, 115
  oxidation states of, 103
  in plant growth, 158–159

Nitrogen-14, abundance of, 109
Nitrogen balance, 181–182
Nitrogen cycle, 103
  biological, 114
  biological *vs.* geological components,
    176–177
  energy sources, 176
  and fertilizer efficiency, 165–166
  geological part of, 114, 116
  plant growth, role in, 159–160
Nitrogen fixation, 159, 176–177
Nitrogenous fertilizer, 42
Northern Hemisphere grasslands, 88
Nuclear binding energy, 106
Nuclear fusion, 105–106
  hydrogen fusion, 108–109
  in massive stars, 109
Nutrient cycling in ocean basins,
  179
Nutrient pollution, 149

Ocean–atmosphere carbon fluxes,
  196–197
Oceanic deserts, 195
Optimal grazing, 98
Oreskes, Naomi, 27

"Paleo" diet, 35–38
  benefits of, 37
  nutritional value of, 37
  Pleistocene diet, 36
Paris Agreement climate goals, 84
Pelletier, Nathan, 18
Phosphorus
  chemical flexibility of, 102
  in DNA and energy metabolism, 101
  dynamics in small pond, 185–186
  nucleosynthesis of, 105
  oxidation states of, 103
Pith, 201–203
  benefits of, 202–203
  bitterness of, 202
  saving, 201–202

Plant-based diets
conventional diets *vs.*, 35
coronary heart disease mortality, 32
effects of, 23
protective nutrients from, 21
Plant eaters, 34, 36
health outcomes of, 30–31
nutritional adequacy, 33–34
nutritional yeast for, 201
Plate tectonics, 177–178
Pool–flux model, 181–194
anthropogenic climate change and
Earth processes, 193–194
anthropogenic flux, 186–187
cyclic nature of Earth's attributes,
183–185
dead zone, 191–192
defining, 181
fertilization, 190–191
irrigation, 189–190
net water vapor flux, 187–188
nitrogen balance, 181–182
phosphorus dynamics in small pond,
185–186
Poore, J., 68, 69, 74
Potassium, 205
in electrochemical balance and nerve
function, 101
radioactive, 180
Poultry
beef *vs.*, 12–15
Processed precooked food, 29
Protons, 103–110
elements with odd number of, 110
fusion of, 105

Reforestation, 78–83, 95, 78–79, 81, 83, 95
Reforesting croplands, 82
Regenerative grazing, 97
Relative humidity, 125–128
drying times, 124
expression for, 125
suppression by subsidence, 126, 128

Rewilded land, 81–82
biodiversity benefits of, 81
growing alternative food on,
82–85
reduced food yield from, 82
Rewilding rangelands, 82
Richter, Conrad, 26
Rinella, Steve, 37–38
Rockies, 137–138
Ruddiman, Bill, 3–4, 5
Ruddiman Hypothesis, 3, 5–6
Ruminant digestive system, 134–135
Ruminant grazing, 42. *See also* Cattle
grazing
croplands for, 93
grasslands for, 88

Salad preparation, 210–212
finocchio salad, 211, 212–214
nutritional benefits of, 212
Sargasso Sea, 195
Savory, Allan, 89
Sea ice-albedo feedback, 60
Sea ice feedback, 61
Sea of Grass (Richter), 26
Semi-arid grasslands, 81
Semi-lush lands, in Texas, 79–80
Shepon, Alon, 21
Soil
agriculture, role in, 164–165
buffering climate extremes, role in,
165–166
composition and properties affecting
crops, 165
degradation, agricultural pollution
and, 166–168
fertility, role of organic matter in,
168–169
mineral-rich, impact of ice ages on,
154–155
mismanagement, 167–168
moisture and productivity, impact of
climate variability on, 164–165

as nutrient and water reservoir, 160
planetary origins of, 154
porosity and water retention, 168–169
Soil carbon storage, 47
  Aryal synthesis, 54
  cattle grazing, impact of, 49, 54–55, 65
  grazing sequestration enhancement,
    66
  of rangelands, 70
Soil conservation issues, 92–93
  adaptive multi-paddock grazing, 93–97
  mob grazing, 98
  optimal grazing, 98
  regenerative grazing, 97
  suggested beef grazing practices, 97–98
Soil erosion, 11–12
Solar energy, 173–174
Stellar life cycles, 107
Subduction zones, 114–116, 178–179
Subsidence (atmospheric)
  air temperature and humidity, impact
    on, 128–129
  apple farming, effects on, 143–144
  boundary layer effects of, 129–130
  Clausius–Clapeyron relation, 127
  compressional warming induced by,
    139
  and dairy farming, 121–130
  lettuce-caterpillars analogy for
    dehydration, 138–139
  near surface evaporation and, 137
  over Great Plains, 138
  regional drying, impact on, 138
  relative humidity and, 125–127, 139,
    144
  surface air properties, impact on,
    124–126
Supermarkets, 201
Surface temperature, 111
Sustainability
  carbon reductions, policy challenges
    in, 84
  environmental, grass-fed beef and, 85

voluntary dietary shifts *vs.*
    bureaucratic climate policies, 84
Sustainable intensification, 207–208
Swaminathan, Akshay, 21
Sweating, 123–133, 136, 137
  efficient, relocating dairy cattle for,
    141
  induced heat loss, 130
  productive, 124
Sweet cherry farming, 145

Teague, Richard, 89, 90
Texas ranch, 79–80

Unprocessed red meat consumption, 29,
    31, 32
Uranium, 102
USDA Conservation Reserve Program,
    81

Vallis, Geoff, 138

Wang, X. 34
Water pollution
  from fertilizers, 166
  nutrient runoff and, 162
  regionalized livestock production and,
    149
Water retention, impact of
  gravitational field on, 152–153
Wegener, Alfred, 27
Westerlies
  apple farming, role in, 143
  interaction with Rockies, 137–138
Westward Expansion, 26
Wheat, land allocation to, 146
Whole Foods, 199
Wildlife harboring potential, human
  impact on, 82
Woods Hole Oceanographic Institution,
    194

Publisher contact:
The MIT Press
Massachusetts Institute of Technology
77 Massachusetts Avenue, Cambridge, MA 02139
mitpress.mit.edu

EU Authorised Representative:
Easy Access System Europe, Mustamäe tee 50, 10621 Tallinn, Estonia
gpsr.requests@easproject.com

Printed by Integrated Books International, United States of America